THE SAND SHEET

THE SEVENTH GENERATION

Survival, Sustainability, Sustenance in a New Nature

M. Jimmie Killingsworth, Series Editor

A WARDLAW BOOK

THE SAND SHEET

ARTURO LONGORIA

Foreword by M. Jimmie Killingsworth

TEXAS A&M UNIVERSITY PRESS • COLLEGE STATION, TEXAS

Copyright © 2017 by Arturo Longoria
All rights reserved
First edition

This paper meets the requirements
of ANSI/NISO Z39.48-1992 (Permanence of Paper).
Binding materials have been chosen for durability.
Manufactured in the United States of America

LIBRARY OF CONGRESS CATALOGING-IN-PUBLICATION DATA

Names: Longoria, Arturo, 1948– author.
Title: The Sand Sheet / Arturo Longoria.
Other titles: Seventh generation (Series)
Description: First edition. | College Station: Texas A&M University Press,
  [2017] | Series: The seventh generation: survival, sustainability,
  sustenance in a new nature | Includes bibliographical references and index.
Identifiers: LCCN 2016024642 (print) | LCCN 2016037163 (ebook) |
  ISBN 9781623495008 (pbk.: alk. paper) | ISBN 9781623495015 (ebook)
Subjects: LCSH: South Texas Sand Sheet (Tex.) | Natural history—Texas,
  South. | Xeric ecology—Texas, South. | Shrubland ecology—Texas, South. |
  Coastal plains—Texas, South. | Longoria, Arturo, 1948—Homes and
  haunts—Texas, South. | Texas, South—Social conditions—21st century.
Classification: LCC QH105.T4 L66 2017 (print) | LCC QH105.T4 (ebook) |
  DDC 577.309764—dc23
LC record available at https://lccn.loc.gov/2016024642

FOR KERMIT SPEEG

Contents

Foreword, by M. Jimmie Killingsworth	ix
Introduction	1
Chapter 1	3
Chapter 2	19
Chapter 3	27
Chapter 4	33
Chapter 5	39
Chapter 6	45
Chapter 7	55
Chapter 8	61
Chapter 9	65
Chapter 10	71
Chapter 11	81
Chapter 12	91
Chapter 13	103
Chapter 14	109
Chapter 15	119
Chapter 16	135
Notes	141
Glossary	145
Index	147

Foreword

Arturo Longoria's time-tempered wisdom makes this book a joy to read and a valued set of lessons. The topics range from how to make a life in near-uninhabitable places to encounters with "long distance travelers" (ordinary desperate folks traversing *la frontera* and modern desperados alike). Illumination arises from Longoria's familiarity with and understanding of the flora and fauna of this little-known but fascinating region in South Texas, his deep knowledge of the place's human and natural history, his experience of the wild, his sensitivity to finding just the right language (honed in the writing of several books and kept up in his outstanding blog "Woods Roamer"), the honest love of the land and the virtually jealous protectiveness he feels toward his place on the earth, and his skill as storyteller. The rare combination of these elements generates the kind of writing and the human spirit that we celebrate in the Seventh Generation series. Add the intensity that comes from surviving a near-fatal illness and the remarkable woodcraft and skills related to survival and sustenance in an endangered and endangering land, and the fit for the series is perfect. More important, it's the kind of book that demands respect—for the author certainly, but also for the land and the values the author represents so effectively. When he tells the oil drillers in one episode of the story, "We will be respected," the reader feels the heroism of the individual and the people who stand with him on the land they love.

Just a word on what I mean by tempered wisdom: Longoria writes so beautifully of what it means to be native in a place of constant change. Some introduced species, such as the reeds that came with the Spanish colonialists and proved so useful to the native peoples, are as much gifts as they are intrusions on the native flora, while other introduced species (not to mention industrial practices such as oil exploration and fracking) appear at times to be curses on the land. Things left behind by even the most careless intruders, bits of metal and other refuse, nevertheless have their uses in the repurposing labors of survivors like Arturo Longoria who make such things as knives from old files to carved spoons and bowls for daily use. Longoria gave up hunting years ago, but he still makes bows and arrows to keep the craft alive and satisfy the spirit of survival that he nurtures.

His tempered wisdom also informs the attitude he exhibits toward the travelers from the south, who range from poor folks seeking a better life to criminals plying their trade and the "coyotes" who guide the people and often abandon them to die of thirst or starvation on this range of American desert. To Longoria, they are all people, individuals, who deserve his cautious attention. He never sentimentalizes them; he never demonizes them. He scrutinizes their faces, their clothes, their speech, their many purposes—sometimes with a weapon in his hand, but always with water and food at the ready when the occasion demands. Likewise, with the border patrol: the various agents earn his respect and friendship, or his contempt, depending on their actions and attitudes.

This book demands attention and respect. If Thoreau had grown up in South Texas and outlived his own tuberculosis to arrive at the state of tempered wisdom, he might have written *The Sand Sheet*. It's that good.

M. Jimmie Killingsworth
Series Editor

THE SAND SHEET

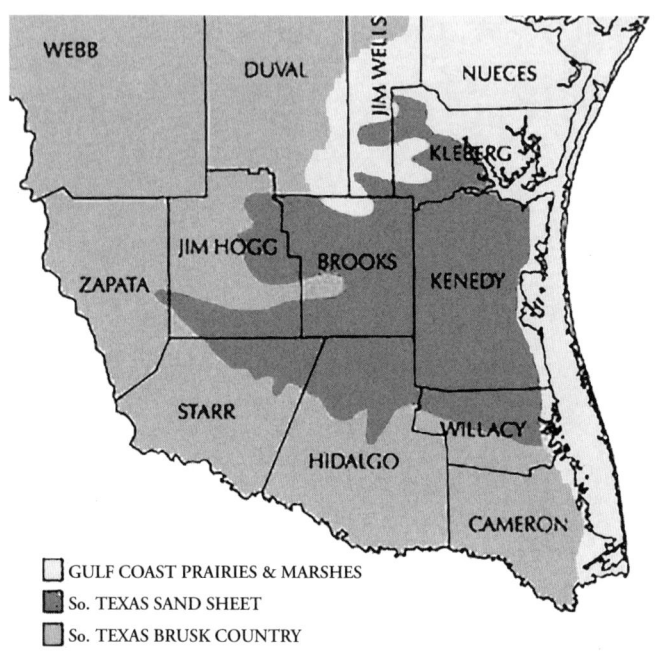

Introduction

The South Texas Sand Sheet is one of the largest and least understood regions in Texas. A Holocene phenomenon estimated at nearly 12,000 years old and larger than the states of Delaware and Rhode Island combined, it arose when the sands from an ancient sea shelf began blowing inland at the end of the Pleistocene. As the sands swept northwestward great dunes formed, and over time these eolian drifts extended into what is now known as Deep South Texas. Although a scattering of scientific papers examining the geology, archaeology, and ecology of the region have been authored, no text has been written about living on the Sand Sheet as it exists now and as it did in former times. This is the story about my life on these sands as a naturalist and writer and how I came to live in this remote region. *The Sand Sheet* depicts a place that is unknown to most and yet as of late has made national news as hundreds of immigrants have perished in what seems for many an interminable desert. My quest to know the hardwoods used by prehistoric people to make bows in the Coahuiltecan Geographical Region is also part of this story, as are my attempts to better understand the primitive skills employed by the Neolithic people who dwelled near what was considered a no-man's land. Combining my love for nature and bushcraft, I have endeavored to show the connection between the two as well as explain the intricate links ancient people developed toward the land as it supplied their needs. I have included a glossary at the end of the story with the scientific names of the plants mentioned in the text. A map of the region is printed on the inside cover for orientation purposes. Random photos of the Sand Sheet are also included. Since I am a knife and bow maker I have also added photographs of some of the knives and selfbows I use in South Texas bushcraft.

ARTURO LONGORIA

1

My youngest son was nineteen at the time. What if he was alone, panicked, and thirsty and had been robbed and was within hours of suffering a slow and terrible death? *What if,* I kept wondering?

"There's no water for over fifty miles," I said.

The boy looked at me but said nothing. He was nineteen and tall with hazel eyes and unblemished skin.

"What happened?" I asked.

"They left us and took our money," he muttered.

I nodded but did not ask him more. Besides, I knew the story. Repeated over and again it echoes across these barren lands like the daytime calls of ghost doves and nightly hoots of horned owls. "Drink slowly," I said, after someone brought a glass of water and placed it on the table between us. The boy sipped the water even as his hands trembled, but mostly his eyes stayed fixed on the weathered planks of the old porch's wooden floor. Yes, I knew the story. Along the southern banks of the Rio Grande people gather in makeshift camps and talk of the great South Texas desert. If you become lost you will surely die. It lies north of cities choked with their indifferences and anonymities and beyond Third World *colonias* stretching along the skyline. The desert does not appear from a distance but instead becomes abruptly real. As flat as a cast iron *comal* its amoebous sands spill outward from the Gulf of Mexico devouring Kenedy County, most of Kleberg and Brooks Counties, and the northern third of Willacy and Hidalgo Counties and sends vast pseudopods into Starr, Jim Hogg, Zapata, and Jim Wells. At 2.2 million acres it exceeds the size of Delaware by an additional half-million acres.

Interrupted occasionally by great dunes and cavernous gullies and with nary a natural pond, stream, or even muddy puddle from which to drink, the only established water holes are those fed from wells dug to sate thirsty cattle.

"How will I be able to sleep knowing I let you go into that godforsaken place," I said. "If perchance you find water, it will be salty."

"I wanted to see my father and brother," the boy said.

"Where are they?" I asked.

"New York City."

I was quiet a moment trying to grasp the enormity of his quest. Although at the same time I wondered: Who is this boy's father who allows his son to come so far and risk so much?

"Do you know the distance from here to New York City?" I asked.

"No, but if I can make it to Houston I will get a bus."

"Houston is far away."

"But they said Houston was that light."

Ah yes, the light. Glowing in the night along the horizon to the northeast nothing but a recharge station from which gas pipelines converge and then go onward like great worms slithering beneath the soil.

"That's not Houston," I muttered though I had heard that story as well. "We will rest here for a few hours," they tell them. "See that glow to the north. That's Houston."

The look in his eyes said he did not understand, so I called for someone to bring a map. "This is where we are—" I pointed to a spot at the edge of the vast sand sheet then dragged my finger north by northeast across a page marked with red lines and black letters "—and this is Houston way over here." I flipped to another page with a map of the United States. "We are here and New York City is there."

He shook his head as if the meaning of what I had just told him was slowly beginning to take form.

"It's over three hundred miles to Houston and nearly two thousand miles to New York City," I said. "But if you go north from here you will be dead within three days."

I had been here before. Not at this very spot but in places alike in other ways. They are abandoned by local smugglers after arriving from El Salvador and Guatemala and Brazil and China and Ireland and Russia and from around the world. They journeyed first on trains or airplanes and then buses and now on foot. They came for work or as with the boy to join relatives living in that veiled underground spanning from coast to coast and as far away as the Canadian border.

"You said your father is in New York City?" I asked if only to fill the silence. "Where are you from?"

A sandwich was placed on the table next to us but the boy was too nervous to eat. "Go slowly," I said. He breathed deeply and then glanced southward as if reaching out to something in the far distance. "El Salvador," he mumbled, his voice a melding of exhaustion and sadness. "My mother and younger brother live there."

I watched as he sipped his water and tried to eat, though I could tell he was merely being polite. "It's all right," I said. "No one is going to hurt you."

I kept telling myself I could not allow him to cross the desert. Somehow I must find a way to stop this before it was too late. At last I said, "I will not tell you what to do. But you must know that you'll not survive the sands. You do not know this land. You'll be lost without water and with no one to help you." All the while I kept seeing my own son wandering alone and scared with nothing to eat or drink. And in an instant it was no longer an abstraction, a story told on the nightly news.

How long we sat there I do not recall, but surely it was only a few minutes. I had spent a torturously hot day working on a house set within a stand of granjeno, mesquite, and brasil not far away. We'd run water from a well to the site and a few days before had connected electricity from the main power line to our meter pole. A sinuous ranch road winds south near the cabin parallelling an old fence and traversing three locked gates each a mile from the other until it reaches the hamlet of San Isidro four miles to the south. My mother attended school there decades ago, and in fact, she grew up seven miles away at a place my grandfather christened El Centro, Texas, in the early 1920s.

The boy remained quiet as he looked off into the distance. But when he glanced at me I saw tears in his eyes. Reflected in those tears, as if images of places faraway and dreams unfulfilled, I glimpsed a mind acknowledging what could not be made so even as the heart yearned otherwise. A reluctant bite from his sandwich and a sip of water; then he sighed and in a low voice said, "Call them."

My mother told me stories when I was a child about the place my grandfather named El Centro. I learned how my mother and her siblings spent their youth in unison with the land. There were stories about exploring the thick woods surrounding their house, but also how one brother was bitten by a rattlesnake and another brother severely burned after attempting to smoke out a rattler with gasoline. My life too has been spent close to the land. I lived in relative isolation along a large lake for nearly three years and then in a cabin hidden by dense woods for almost four more years. Then I built another cabin on a hilltop called Mira Lejos. Alas, to return to a place where people laid waste what bits of nature surrounded them. One day an early sign of winter swept over me, and I hovered above that great abyss into which all men eventually fall. I had been feeling weak but figured that perhaps I was dehydrated after working in the hot sun a few days before. I'd noticed my ankles swollen for about a month, and I'd developed a severe rash on my lower legs as a result of a foray into the woods several months earlier. During that trip I'd stepped into a nest of pin-point-sized ticks or what people along the borderlands call *pinolios*. Hundreds of them scaled my legs from ankles to knees and then took hold. By the time I discovered them, hours had passed. It took three more hours to dislodge the ticks using a magnifying glass and a pair of extra fine tweezers. But it was too late. The damage had been done, and on that afternoon as I sat in the shade of a palo blanco tree someone walked up to me and said, "Your eyes are yellow." A score of hospital stays followed, but no one knew what was happening. The gastroenterologist I was referred to would shake his head and utter phrases like "that's impressive" when he read my test results. Liver enzymes high, bilirubin high, platelets low, and the news kept getting worse. A failed endoscopic retrograde cholangiopancreatogram (ERCP), an excruciatingly painful percutaneous transhepatic cholangiography (PTC), a liver biopsy that began in my jugular vein, crossed through my heart, went down my inferior vena cava, and then entered my portal vein. During the biopsy my heart rate fell precipitously and I nearly died on the spot. There was a plasma transfusion, a couple of CT scans and an MRI. I had reached a point where I could barely walk, and most of my days were spent sleeping. At last the local specialist came to terms with his ineptness and I was referred to the Transplant Center at the University of Texas Health Science Center in San Antonio. My MELD Score (Model for End-Stage Liver Disease) was high enough to put me on the Texas Liver Transplant List. So

I prepared myself for what seemed the inevitable. Except perhaps that I am stubborn and dogged. One evening I decided to take out the treadmill and simply walk myself into the ground no matter how miserable I felt. Despite my condition I was still eating well, and though relatives would look at me and then cast their eyes at the ground, I reasoned the worst thing that might happen would be to drop dead on the spot, which was not so terrible given my condition. I climbed onto the machine and took hold of the handrail. Then as best I could I began to walk. Gradually I cranked up the speed even as my legs became jelly and my balance not unlike a drunk negotiating his way through a dark alley. But it was *not* suicide. On the contrary, it was tenacity refined and pure, a refusal to lie on the couch and simply waste away. Even so, as I turned up the speed I expected to drop and be dead before I hit the floor. At the end of eight minutes, soaked in sweat and feeling giddy, I turned down the knob until the treadmill trudged to a tranquil stop. I stood a moment then stepped down and began walking slowly to the bedroom where I removed my clothes and then managed a shower. Afterward I sat on the recliner feeling surprisingly odd. In an ambiguous way I felt better though I was still very sick. I'd become a vegetarian, following a regimen in a book called *The Liver Cleansing Diet*, so I ate a salad then took the medicines I'd finally been prescribed then went to bed. Day after day the routine remained the same: Walk the treadmill, eat a vegetarian meal, take my medicines, and then rest. Fourteen days went by and it was time for me to go to San Antonio.

On that first visit I was escorted into a lab where a phlebotomist took nearly two dozen vials of blood. I was scared and depressed. This was how it was going to end. That afternoon I had an appointment on the eighth floor with a man named Dr. Kermit Speeg. A degree in medicine from UT Southwestern Medical School in Dallas and a PhD from Rice University. When he entered the room there was no fanfare, though I'm not sure why I would have expected there might be. I smiled but I suspect he saw apprehension in my eyes. In his hands a nondescript folder, brown or blue, I don't remember. The analysis of Arturo Longoria: a man of the woods, a nature lover, an amalgam of environmental advocate and primitive skills devotee. The sick one. The dying one? The folder would reveal the facts. But I do not remember a word he said at that first meeting. I recall his smile and perhaps a mention that another ERCP was going to be required. I did not know and perhaps neither did Dr. Speeg that a fork had been reached in the road marking the life of this man, Longoria. Had I continued seeing a doctor who was either incompetent or indifferent or both I would not be here writing these notes. But instead I'd journeyed to San Antonio and it mattered little to me its claims about the Alamo or the Spurs or the River Walk. As far as I was concerned it was the home of one Kermit Speeg and he would see me through. At first my visits were every month. I had another ERCP and this time it was successful, and I have a young doctor named Hooks to thank for that. Many more tests with seemingly countless results followed. Weeks turned into months. I think of it now like a bird persevering through a hailstorm. Progress was slow and not always positive. My tests wandered from getting better to not so good to back on the right track. Then one spring afternoon I decided to pump up the tires of my old bicycle and head off to a wooded trail. The treadmill had not killed me and I had come to the point where I was walking two miles three times a week. My new medicines were having a positive effect as well, so maybe I could take my chances on the ponderous balloon tire bike. I placed the bicycle in the back of my pickup and headed to a ten-mile trail in a neighboring town. But while everyone else was riding multigeared cycles, I was determined to ride my ancient one-speed cruiser. I began visiting the bike trail three times a week and taking long walks in between. Then one late summer's day a year after my first visit with Dr. Speeg I realized I felt normal or practically so. It was an insidious thing not unlike waking up after many sleepless nights and realizing you finally had a full night's rest. Even so, I feared a false springtime with bitter snows returning. So like a man newly converted to some religious order I kept to my diet, exercise, and medicines. For the thought remained: What is the worst thing that can happen but to collapse and die and thus save myself the misery of a slow wasting away.

A few years previous I'd watched that very thing happen to my father. One day he was sitting at the dinner table in conversation with my mom when she noticed he began using his fingers instead of his fork. His speech became slurred and nonsensical. Mom called my dad's doctor and he suggested my dad go to the hospital immediately. My mom, however, did what many people do when facing imminent crises with their loved ones. She panicked. Instead of summoning an ambulance she looked for me and finally located me at the dentist's office. I remember the office receptionist walking into the room and saying, "You need to call your parent's house right away."

I found my mother somewhat incoherent. "When you have time could you come by the house and take your dad to the emergency room. I called his doctor and he thinks he might be having a stroke.

"What?"

"A stroke," she repeated.

"Mom, call an ambulance right now. Do you understand?"

"Well, you don't have to get so touchy about it."

"Mom, please do as I say. Call an ambulance."

When I arrived at my parent's house I saw the ambulance parked in front. I remember the look on Dad's face as they wheeled him out on a gurney. He smiled when he saw me. But I was too worried to act wisely and said, "Do you know who I am?" I could see him struggling for the words and I recall the frustration on his face as he tried to speak. But he could not talk, and in the two months that followed I watched my father suffer an agonizing death. He never went back home though for one brief moment and for reasons unknown he regained his speech one sunny morning. As my mother held the phone he spoke to me saying all was fine. "Nothing to worry about," he said. I would like to think he had a smile on his face when he spoke those words. But sadly, it did not last. And on the day he died I sat there in the hospital room listening to his tormented breathing, his hopeless gasps for air, his lungs and heart failing by the minute. It came on so quickly, those last few moments, that I can barely remember them though I have tried over and again. He simply grew very quiet and a blanched look swept over him. He was a man with elegant green eyes, tall with a commanding presence. He had owned brick plants in Mexico and thus made his fortunes in life. Now he appeared frail and small.

A nurse entered the room and said, "There's still a pulse." A minute later another nurse entered and said, "The EKG is flat. It's over."

I phoned my mother and said, "He's gone, mom."

"He's gone?" she asked, her voice in a vacuum lost and frightened.

I never really knew my father, for he was not a man inclined toward family. But I loved him and cherished every moment I had with him. I know however that any man who could endure what he endured is brave beyond words. And I am not that man. So I worked my bicycle's pedals, plodded the treadmill, and took long walks until at last I knew the time had come to return to the woods. I'd stayed in a city for too long and wanted nothing more of it. My wife, Norma, said she'd go with me as she'd done in the past. In the early years we lived along the shores of Falcon Lake in Zapata County. Not long afterward we built a cabin in the woods and while there Matthew and Ethan came into this world. At last I was heading home.

The house was built and what plagues we endured in the process were survived. You see, there are those who belong in nature and those who should stay away. A man born to the wilds cannot live in a city; likewise, a man of the city will always be unhappy in the deep woods. Even so, I was not all that sure about the location we'd chosen. I'd known the Sand Sheet as a boy, having hunted deer on various ranches, and I'd even owned a parcel of land on the Sheet in the early 1980s. When my older boys, Nomar and Jason, were little we'd camp at the fifty-acre place we called Rancho Escondido. In the northeastern corner of Hidalgo County, it was my first real encounter with the sands. Even so, in order to know the Sand Sheet one must live there, and by that I mean residing intimately with the land and not in some opulent enclosure that holds its tenants at a distance. Nomar and Jason are grown now and have their own families, but I'll not forget wrapping their ankles with tape to keep the burrs from scratching them as we walked the woods. Grass burrs and sand, that's what I recalled. So when we found the place in northeastern Starr County I told Norma I had reservations about living on the edges of the desert.

"It's a hard place to be," I said.

She smiled. "We'll fix it up you'll see."

At our house we were in the midst of the western half of the South Texas Sand Sheet where clumps of granjeno, brasil, and mesquite grow in what are known as motts. To the east along

Highway 281 and onward toward the coast are stands of live oak, but in the western part of what locals call *el desierto* oak is mostly absent. Gathered within these motts are also stands of lotebush, colima, and Texas lantana. But the Sand Sheet is marked not so much by what lies within the motts as what grows between them; for in truth, the South Texas Sand Sheet is a world of herbaceous shrubs and grasses. When rains come the sands are enveloped by a kaleidoscope of colors. Yellow, red, blue, and violet mix with white and orange to form hues that meld as breezes blow floral clumps one against the other. Grasses form mats across the land, some sprawling close to the earth and others swaying gently like wheat in the wind. But walk five hundred yards south of our home and the scene changes as if stepping through a cosmic wormhole only to emerge in some distant universe. The deep eolian sands disappear and instead one finds a soil richer in clay and iron and more solid to the touch. Gone too are the motts and herbaceous shrubs in between. While the Sand Sheet is a young system evolving toward increased plant diversity, the land to the south of the cabin is ancient thorn forest with scores of hardwoods and an abundance of cacti. On the Sand Sheet flowers and grasses live but a few days or until desiccation leaches their colors away and the land becomes a brown mono-scape with only the hardiest woody shrubs surviving. Mesquite, granjeno, and brasil—the primary hardwoods on the western part of the Sand Sheet—muster the energy to stay green and produce fruit through long periods of drought. Set on this biotic borderland, the walls, porches, and metal roof of our home resided between the thick brush and the sandy desert. In spots nearby the boundary ran east and west while in other locales it traversed north and south. Like a young river meandering overland the line separating sand sheet from thorn forest wound haphazardly, swooping one way and then another. West of the house the boundary swept gently to the north. To the east the border swayed and quivered in an unsteady gait lurching to the south and then slogging farther eastward.

The Sand Sheet owns its own mysteries hidden beneath dunes and flats that overlay an older geological stratum known as the Goliad Formation where 12,000 years ago Native Americans made their lives. From the Gulf of Mexico, the sands rise gradually to about 400 feet above sea level at the Sheet's far western edges. Dunes inch across the flats like giant slugs fanning outward gobbling up trees and shrubs and in their wake revealing earth not seen for over ten millennia where atlatl points, as if abruptly resurrected, squint at a sky concealed for what seemed an eternity. Once there were streams and ponds and even lakes atop this ground, but when the sands blew inland from the coast the people fled, and the streams, rivers, and lakes were erased as if they had never been.

Utility poles brought electricity to our little house. A well delivered reasonably good water though we purchased cleaner drinking water in town. At night pauraques and coyotes whistled and yodeled above a backdrop of harmonizing crickets. I think of it as nature's white noise. In front of the house three faucets dripped water into clay bowls placed near birdfeeders. Inside, binoculars dangled within easy reach, and we were forever looking out the windows to spot new birds as well as enjoying those that frequented the feeders and watering stations. Distant cities to the south and a highway far to the east existed as if on another planet.

I think back on those first months of building. Everything we brought got stuck. The truck arriving from McCoy's in Rio Grande City forty miles to the southwest sank into the sand and we had to borrow a large tractor to pull it out. The truck carrying caliche for the driveway spun down nearly to its rear axle, and it took an hour to dig it free. And the truck from Magic Valley Electric plunged down—its tires like angle grinders burrowing into balsawood—until the universal joint rested firmly on the sand. They had to bring another larger truck to pull out the first truck and the second truck also got stuck. The work crew was ready to call it quits, but I desperately needed electricity in order to start construction. So I loaded all seven men into my pickup truck and then treated them to lunch at the 1017 Restaurant in San Isidro. It was a hundred and thirty-six dollars well spent. When the crew returned to the cabin site both trucks had been freed and so they finished the work. Regardless, it was that moment of connection to the electrical pole on the narrow trail leading to the house that marks, at least in my mind, the beginning of life along the Sand Sheet. Plunging a posthole digger into the soft sand deeper and deeper and then anchoring the meter pole into the ground. Then one of the men from Magic Valley Electric climbed the pole and mated the heavy drop wire to the galvanized weatherhead. A few additional connections and a quick overview and he glanced down at me and said, "You've got power."

Does a building project ever go smoothly? In this case we faced not only the logistics of hauling supplies to a secluded outpost but also dealing with sellers who had, unbeknownst to us, incurred a sizeable lien with the IRS. Muddled in their own disputes it fell upon our shoulders to obtain a partial release of that lien so we could proceed with acquiring the property. There were indeed times when it seemed all was lost. But I have, if nothing more, acquired a persistent patience in my life and thus managed to purchase the land and finish the job.

"Call them," the young boy repeated. When the Border Patrol vehicle arrived and the boy was about to sit in the backseat he reached out and hugged me. "You'll be okay," I said turning to the two agents with a warning, "Take care of him." They too were young men and looked as out of place in this land as the boy who now sat in the vehicle drained and demoralized and perhaps thinking of his father and brother in New York City and his mother and younger brother back in El Salvador.

"Are you seeing a lot of people coming through here?" one of the young agents asked.

"Now and then," I said. "Mostly they're lost and thirsty."

2

I'd sent someone to open the three gates leading back into the world. As the Border Patrol truck rounded the curve and drove out of sight I looked north, contemplating that great expanse of waterless sand stretching nearly sixty miles. A young man who's lived in the area since he was a boy tells the story of how he and his older brother and a companion crossed the desert when he was only eleven years old. He helped in the construction of the house, and one afternoon as we sat in the shade of a mesquite tree he told me his story. His name is Amador.

"We walked at night," he said. "Each carried a gallon of water, two small cans of Vienna sausages, and a can of beans. We crossed the Rio Grande south of Starr County and headed north. I was eleven, my brother was twenty-two, and his friend was about thirty. But the days were hot and by the end of the first night and day we'd finished our water. The sausages and beans were salty and that made us even thirstier, but we needed salt almost as much as we needed water. It took us three days to reach the desert walking only at night and then hiding and trying to sleep when the sun was out. South of the desert, walking through thick brush, we'd find windmills and refill our jugs that we'd wrapped in cloth so the thorns wouldn't pierce the plastic. The water we found was always sweet, and we had no great problems other than we'd eaten all our food by the third day. We walked about fifteen miles each night, and when we reached the desert we found a clump of trees and then hid as the sun came up. We'd not thought of water because we figured there'd be wells where we could refill our jugs. But on that first day in the desert as we rested under the shade of a cluster of granjeno and mesquite trees the winds began blowing. By noon our faces and necks and even our hands were raw from the biting sand. We'd kept a piece of plastic from when we'd crossed the Rio Grande and we tore it into three parts and used the pieces to cover our faces. But by midday the sand had nearly buried us. There were stands of trees here and there and in between nothing but dried weeds. By nightfall we felt weak from thirst, and the blowing sand made our vision blurry, but there was no water to wash out our eyes. So we just kept walking. It wasn't until about midnight that the air began to cool.

"At sunrise we saw a windmill in the distance. We'd gone over a day without water and we walked as fast as we could, but when we got to the well we found the water too salty to drink. Our mouths felt as if they were full of sand. We'd seen no footprints or even tire tracks near the pond, and we were too exhausted to go on so we decided to sleep there by the water we could not drink. My brother decided to look around and he found a clump of nopal near a stand of brasil trees. He'd brought along a small knife and with that he cut several pads and removed the spines and we ate the nopal raw. We'd been warned not to bring any sort of knife in case the Border Patrol caught us, but had it not been for that little knife we would not have been able to eat the cactus. By nightfall we were burning with thirst. And though the nopal had helped we were still hungry. My brother's friend wondered aloud whether we could walk another fifteen or twenty miles having had no water and little food in two days. But we had no choice, so at sunset we started out again. Sometime around midnight we heard a helicopter to the south but we never saw it. We knew it was better to travel at night when it was cooler and there was less chance of being seen, but that was also when rattlesnakes were out and so we walked in fear. I'd never seen snakes like the ones in the desert, irritable and as long as a man is tall. By morning we were mad with thirst. My brother's friend began crying, saying he was too old to cross the desert. He said if he didn't make it then to please tell his family that he loved them. But my brother would hear none of it. He told him if he did not compose himself we would leave him. The next morning, we looked for nopal but found none. I saw a tall grass that resembled wheat and so I pulled off the seeds and tried to eat them. But I did not have enough saliva to swallow so as best I could I spit them out. We had been wearing caps to keep the sun off our heads but we'd dropped the caps in the thickets and were too weak to bend down and pick them up.

"On the sixth night of walking after crossing the Rio Grande and having gone without water for almost three days we reached a river. It was the Nueces. The desert was behind us. We were horribly sunburned and the sand had scratched our faces and necks to the point they were raw and bleeding. My brother was walking like a drunken man and his friend was seeing things that weren't there. I felt as if my body was dying from deep within me. We stumbled down to the water and began drinking and for the rest of the night we slept there along the bank. At sunrise, having refilled our plastic jugs, we walked up the bank hoping we were close to a town where my brother's friend said he had an uncle working on a farm. But as we stepped out of the woods we encountered a man in a truck. He looked at us and we could tell this was bad so we ran back down to the river and hid in the trees. Within a few minutes a Border Patrol helicopter arrived and started flying overhead. They spotted us and soon we were surrounded by several men. They captured us and within a day we were at the port of entry in Laredo where they made us cross back into Mexico. But then eight months later I swam back by myself and ended up here at San Isidro where I've been living and working ever since."

Both Native Americans and Europeans knew to keep clear of

the sand. The roads and trails made by Spanish travelers called *Los Caminos Real* leading northeast out of the interior of Mexico into Central and East Texas circumvented the desert, as did the Comanche traces from the Texas Hill Country into Mexico.[1] Even the adjacent thorn brush west of the Sand Sheet known as *las brasadas* or "burning lands" harbored few permanent settlements because water was scarce and life cruel. But before the sands arrived there were people living on the land. Perhaps the greatest misconception about the earliest humans who dwelled in what is now South Texas is that they vanished. The good news, however, is that the prehistoric settlers of the region and indeed throughout the state still exist in the form of progeny who occupy all levels of the greater community. As has occurred throughout human history people migrated from one place to another, groups displaced other groups, and cultures broadened their influence over vast regions. In fact, throughout Latin America the conquest by Spanish and Roman Catholic authority is clouded as much by the destruction of cultures as of people. Tribes, languages, and mythologies were either erased or severely denuded. Although many individuals survived, their names had been changed and their unique civilizations expunged. This is perhaps the greatest story of lost roots because the narrative plays directly into the lives of those who presently dwell in South Texas and, indeed, across the state. Terms such as Hispanic, Anglo, and Latino may fit political and bureaucratic spectrums but they do nothing to effectively interpret the history of any particular group or place. Regardless, it is the richness of this fusion of muted myths and traditions that becomes truly palpable in settings removed from the rush of modern life. With its low population density and relative isolation from large metropolitan areas, the vast Sand Sheet allows the collective influences of older cultures to blossom, and as such the remoteness and harshness of the land reawakens long-shrouded customs and lore.

When the first people arrived is speculative, but the words *late Pleistocene* are frequently used. At least twelve thousand years ago, though the question remains did they arrive from the west, south, or north? Regardless, they kept near game and always within reach of water. Archaeological digs in Brooks County reveal prehistoric campsites along streams that no longer exist.[2] As Pleistocene ice retreated in the far north and great winds began blowing across the land, the sands off a barren sea shelf swept westward in what must have been a series of epic haboobs that entombed what stories lay beneath.

Throughout the late Pleistocene and into the Holocene ancient people occupied the rich riparian delta to the south along the Rio Grande. Settlements also existed in the north at what is now called Baffin Bay and Corpus Christi Bay. In between along meager streams and ponds lay smaller villages. But as the great desert formed, this middle land began to change. The expanse of sand and undulating dunes created a barren landscape devoid of habitation for man, beast, and plants.[3] Gradually, however, the sands were again reclaimed by the foliage surrounding it, and thus the Sand Sheet is a living laboratory for the study of plant reintroduction and burgeoning diversity.

In the mid-1700s colonists from the Celtic and Basque regions of the northern Iberian Peninsula settled the lands on both sides of the Rio Grande in what are now the Mexican states of Tamaulipas and western Nuevo Leon as well as parts of South Texas.[4] They called it Nuevo Santander after a region in their homeland. Like the ancients before them they stayed close to water, for without water ranches could not flourish and farming was impractical.[5] But to the north the sprawling sand was a seemingly interminable blockade for both men and his beasts, and it remained a no-man's land.

At last our house was completed. But by the summer of 2011 the region from San Antonio to the Rio Grande, spanning approximately 250 miles and known collectively as South Texas, had already experienced months of exceptional drought. In many places lakes went dry and ponds were erased, and rocky scars etched the paths that once marked flowing streams. The land around our house had been heavily grazed and I noted how men's works so easily impact fragile regions. Cows had inadvertently selected for an unpalatable species of croton, and the land to the north and west was dotted by nothing but gray shrubs called silver croton. When north winds blew in late October our porch was covered with sand and I spent a half hour daily sweeping off the concrete blocks that formed the floor of my little shop nearby. Had I made a terrible mistake, I wondered? Was this place simply uninhabitable? And yet others before me had lived on this very spot and to the north stood tiny homesteads where people made their lives. Now we were here and it would have to do. Thankfully, I had Norma who always looked on things in a positive way. On Thanksgiving 2011 we gathered in the little cabin. My two youngest sons, Matthew and Ethan, came to visit. My older sons were far away and not an hour of any day transpired without me thinking about them. Still, the two youngest had arrived from their respective colleges and for that I was deeply grateful. We sat in the small area that constituted the living room and feasted on turkey and dressing. Cranberry sauce, mashed potatoes, string beans, and candied yams. Whole wheat rolls, pecan and pumpkin pie. And lots of freshly brewed coffee. I kept looking at my sons remembering them when they were little and thinking back at the cabin we built at my uncle's place years before. Nothing left of any of that. All gone. The land was sold after my uncle's death and turned into a *colonia* where not a single mesquite tree was allowed to remain. Such was the life I'd come to know. At least we were all doing well and had each other and that was what counted.

Thanksgiving came and went and the boys returned to their colleges. We were once more alone. I'd mope around the cabin missing my children, but Norma would try to distract me. "There are things to do," she'd say. There was a garden to plant, grass to grow to subdue the sand, and there were places to walk and discover. I'd met a few people from San Isidro but mostly kept to myself busy with my labor and living quietly. Of course, that would change over time but for now we'd settled in and life was good.

3

Acquire a chronic illness and then take care of it and you will live a long life or so the saying goes. But in all things there are exceptions and many fall prey to lions hidden in the tall grass. Another saying holds that the sick are the walking wounded. But I believe the truth is perhaps subtler. The sick are no more wounded than all others. They have simply tasted the apple and thus glimpsed their own transience. At least for the moment I had again opened the doors to life and like countless times before stepped forward to roam the quiet woods. Without formality or ritual and with no edict to pay on demand or illicit shame and guilt, the woods ask only to be given the chance to bestow healing and thus reveal the only truth we know to be real. Although the house was built there were still other projects in queue. We needed a storage shed, perhaps two. We lived in a secluded world beyond an asteroid belt of mesquite, granjeno, brasil, and chapote trees. In some places not far away people pushed nature back and so they were exposed to the sun and wind and ultimately the sand. But my philosophy has always been to keep nature close. The trees surrounding the house provided shade and the grass we planted kept the sand at bay. The watering stations attracted birds from miles around, as did a daily ration of grain. I'd made a few forays into the nearby woods but building the cabin had occupied most of my time. Now, however, I could go woods roaming. I strapped my canteen and musette bag over my shoulders and headed through the motts surrounding the cabin then across meadows where grasses and shrubs made resolute efforts to survive despite the ongoing drought. I walked on as if journeying further and further back in time until at last I stepped into a forgotten world.

Imagine a place of industry and commerce abruptly collapsed. Factories deserted, roads disintegrating, sidewalks obscured, and all around steel pipes and giant tanks frozen by corrosion, the metal flaking and discolored. Nothing moves. All is silent. The only footprints are those of coyotes and javelina and a marauding troop of wild hogs. In some places dilapidated dwellings—homes and offices and crumpled sheet-metal sheds—stand choked with rat's nests and black widow's webs. Rotting timbers and the few glass window panes that remain are layered by dust and the discoloration of decades of abandonment. There's an old house surrounded by thorn shrubs and small trees, its roof collapsing, its doors missing. Even from a hundred feet away one hears the drone of a million bees. Some who have neared the building say the hives hang from the ceilings to the floors in nearly every room. In other places cracked foundations etched by the initials of their builders now dead and forgotten appear like australopithecine skulls emerging from the baked soils of the Great Rift Valley. The rear axle of a child's toy, a leaf spring from a 1939 Ford, and piles of old medicine bottles now mostly covered by sand. Giant steel tanks, like Mayan pyramids walled by wilderness, stand corroded and brittle with rickety stairways leading to their tops. Even so, some remember. In the late 1930s and throughout the 1940s and '50s they recall thousands of workers entering with their machines and rigs and then drilling without remorse. The land it seems had no value and was but an obstacle to be scraped and bored into. A man who lives three miles down the road from us remembers that world. His name is Ramiro and he is a direct ancestor of Santiago Peña who founded San Isidro in 1877, its name derived from Saint Isidore of Seville, the patron saint of farmers. As the story goes Santiago Peña found the land poor for farming, sitting as it was at the very edge of a great ocean of sand. When he dug a shallow well his pails lifted water ridden with salt and sulfur. Even so, others migrated to the area and by the turn of the century a scattering of people had joined Santiago. They dug a community well alongside a dense *ramadero* where water flowed beneath the soil en route to the Rio Grande fifty miles to the south. Then they built cisterns with long troughs where cattle watered. Other wells were dug nearby and today those old wells and cisterns sit ghostly and forgotten. A Catholic church was raised a hundred yards or thereabouts from the original well, and in 1904 a post office was established. Even so only a few dozen people lived in the community for nearly thirty years. Other ranch villages emerged not far away, and for decades small *ranchitos* dotted the wilderness for miles in every direction. Delmita and El Centro were both homesteads and family run stores in the 1920s south along ranch road 2294. To the west La Gloria marks the intersection of FM 1017 and FM 755. Santa Elena was a ranch a few miles north of La Gloria along FM 755 that begins in Rio Grande City at the US-Mexico border then follows a northeasterly route until it intersects with US Highway 281 in Brooks County. East of San Isidro where FM 1017 curves abruptly sit a few houses and a dilapidated gin. That place is called La Reforma and my grandmother grew up there a hundred years ago. My great-grandmother, Ines Saenz de Guerra, is buried at a family cemetery nearby.

But Ramiro Peña does not have fond recollections of the days when people from the north swarmed into the area searching for oil: "There were oil rigs for as far as you could see with trucks on the roads all day and night," he said. "I was young and needed a job. But the only jobs available to us were the dirtiest and lowest. The ones who came from the north wouldn't do the hard work themselves because they were soft and lazy. They stood around reading meters and not much more. We hated them. This was our land and they treated us like the dirt at their feet. Finally, I moved to San Antonio where I worked most of my adult life. Later I came back to visit but by then the land had been ruined. It was just the oil they wanted. I knew that in my lifetime it would never fully recover."

They came and plundered and then left. It was as if I was looking at the earth from a thousand years into the future. But instead of atlatl points and stone axes I found deceased drill pipes and wellheads bolted shut and capped, immobilized by the very earth that was pierced and gutted. Like a wound never stitched but left to heal from deep within until at last it seals at the surface leaving an ugly scar. Still, I thought of all those who had been here before me. Those I never knew and those I knew and loved. Likewise, I thought of how the land had changed over time. Then wondered what I would have found had I stood at this very spot a thousand years ago, five thousand years, ten thousand or twenty. I wondered what was it like a million years before and then glanced up at clouds that had shut away the sun. The skies dimmed and the woods before me appeared as if haunted. But then those are my favorite days, not simply because the sun is hidden but because in a paradoxical way I see more clearly the land before me within the darker cast of long shadows. I stepped lightly, maintaining silent footsteps. A small machete called a parang secured in a rawhide sheath dangled from my belt. I'd forged it a few months before from a fourteen-inch mill file. Across my shoulder a two-quart canteen and in my bag a crooked knife, twenty feet of parachute cordage, a ferrocerium rod, and a pair of leather gloves. A wide-brimmed felt hat old and discolored, and in my pants pockets a bandana and a couple of jackknives. I'd come to the point where anything more was like a mat of thorns pressing into my back. I used to carry a rifle but that ended long before I became ill. So on that day I brought a slingshot and a small bag of marbles. When I was young my grandfather and his sons (my two uncles) had slingshot shooting contests. My uncle Bill was a medical technologist and while everyone else used rubber bands made from inner tubes, my uncles and grandfather tied surgical tubing to their slingshots courtesy of the medical lab. We'd hike the woods looking for suitable *orquetas* or forked branches of mesquite or chaparro prieto onto which we fastened the tubing. My grandfather and uncles could easily hit tin cans tossed into the air. I was as enamored with slingshots as they were and yet living in the woods without the use of fancy and expensive equipment came mostly through my love of nature and the desire to exist simply and quietly. In time I learned to make my own wooden spoons, bowls, and cups. I made knives from steel files and leaf springs, and selfbows from saplings and hardwood branches, and I fashioned arrows from the common reed known in South Texas as *carrizo*. I made blowguns from a type of cane (caña) called *Arundo donax*. I learned the art of primitive trapping and added to a lifelong inventory of what native plants are edible or medicinal. I made cordage from agave and yucca fibers and even ortegia. My crooked knife was derived from a six-inch steel file I annealed, reshaped, heat treated, tempered, polished, and sharpened then inletted into a wooden handle secured with cordage. I carried a leather pouch in my musette bag that held my first aid kit.

A mourning dove cooed nearby and I brought cupped hands to my mouth and answered a coo in reply. We talked back and forth as the clouds drifted off to the west and a fan of crimson rays from a setting sun spread over me. At last the dove's coos came no more. I wondered if I might ever hear that same dove again. Probably not, though for a moment we were as one and we said as much if only briefly.

4

At night I'd sometimes see helicopters flying to the west and east of the house. I'd step out on the porch and watch the choppers going back and forth. One night a helicopter flew directly overhead. It was going north but abruptly began circling as a bright beam pierced the blackness from chopper to ground. Norma and I watched from our bedroom window as the beam swept the brush about a quarter mile away. A man I'd met who lives in San Isidro named Tololo Lopez had warned me that sometimes smugglers used the little road leading to the house as a clandestine route between the paved highways to the east and west. "Be careful," he said. "The smugglers can be dangerous." The helicopter flew in a large circle for what seemed almost an hour. I had six blue heelers at the time and before the helicopter arrived they began barking fiercely. I walked out on the porch with a flashlight but saw nothing. A few nights before as I worked in my little shop making a selfbow from an anacua stave I heard a vehicle entering through the closest gate about eight hundred yards southwest of the cabin. Tololo said that keeping that gate unlocked was a bad idea. But a man who lived nearby refused to lock the gate because he hated opening and closing it. Over several weeks I began hearing a vehicle entering through the gate late at night and then heading westward along a narrow ranch road where it would cross through another gate that was presumably an illegal entry into a neighboring property.

I'd often work into the early morning hours enjoying the coolness and penetrating silence around me. I'd hear great horned owls behind the house or screech owls in the deeper woods. Sometimes I'd hear coyotes yodeling far off. But for the most part the quiet was like a woolen cloak draped heavily across the land. If there was a moon, I'd watch it slide through the heavens until it disappeared in the west. On moonless nights I'd look up at the Milky Way and think of it as fine layer of chalk smeared across a blackboard.

In the mornings I'd walk to the gate eight hundred yards away and see tire tracks leading westward on that road that crossed into the other property. Always the same tracks. Then one afternoon the young man named Amador who had helped in the construction of the house dropped by. He'd been staying at a trailer about a half mile to the south. The trailer's owners, Ray and Jovita Vance, had employed Amador to do some cleanup work around their place. I'd given Amador the nickname *Tres Manos* because he could do the work of more than one man. But Amador was apprehensive that late afternoon. "What's wrong?" I asked. He said that during the night a dark-colored van had driven down the road in front of the Vance's trailer— the same little road that traversed the last gate and then neared our house. Ray and Jovita only spent weekends at the trailer and Amador was alone. He said he'd walked out from under the porch where he was sleeping and someone in the van had spotted him. I'd not seen Amador like this. He was scared and wanted to be taken back to the place where he usually stayed. "They threatened me," he said. But he would not elaborate further so we sat quietly a few minutes talking about other things. About a flock of wild turkey he'd seen a couple of days before and a rattlesnake he'd killed with a shovel when it tried to bite him. At last he said, "They're smuggling people and drugs."

Over twenty years before I'd attended a meeting on a private ranch about fifteen miles west of where we now lived. Every sheriff from the entire state of Texas was at the meeting, as were representatives from the Texas DPS, the FBI, US Customs, the Border Patrol, ATF, the DEA, the State Department, and an assistant US Attorney from Washington, DC. The meeting lasted most of the day and chefs were hired to provide a good old Texas barbecue. The year was 1986 if I recall and I'll not forget the talk given by the assistant US Attorney from DC. He warned that an unprecedented crime wave was going to hit the South Texas border. "It might take a few years before it's here in force but it's coming," he said.

When the United States was in a panic after 9/11 another war began to the south in Mexico. While the American president was fixated on a place that had no weapons of mass destruction and was of no imminent threat to the United States, a real and present danger began rising its head below the Rio Grande. A friend of mine who worked as an analyst for the FBI told me one evening shortly after September 11, 2001, that the border was out of control and there was credible information that terrorists were entering the country through sophisticated smuggling networks called cartels. Mexican border towns were abandoned and thousands of people were killed as cartels waged war against each other. Mexican government officials, politicians, bureaucrats, the police, and the army were often in cahoots with these syndicates. I had known Mexico as a boy but now it had become a war zone, a place we dared not venture into because of the extreme violence. "The crimes we'll see along the border in the not too distant future are already happening in places like Mexico City," the assistant US Attorney said at the meeting in 1986. "There'll be kidnappings and home invasions. The drug situation will get much worse." We listened to the man from DC and I think for some it was hard to believe those things would happen on American soil. But after the turn of the new century the violence and crime settled firmly along the South Texas border, and yet the worst was still to come.

"If you want me to take you home then I will," I told

Amador. Later that night I walked to the gate that was always left open and I locked it. The next day I told the man who hated to open and close gates that from now on that gate must remain locked. He didn't like it but I was in no mood to discuss the matter further.

5

I'd taken to using walking sticks to probe the grass and shrubs around me for rattlesnakes. One afternoon about two months after we'd built the cabin I was walking down a cow path and a large rattler crossed but a few feet in front of me. I'd forgotten to bring a stick so I made one on the spot. With my Swiss Army knife's saw blade I cut a straight granjeno branch a bit over four feet long from a stack of poles someone had used to build a small shelter not unlike the wickiups Native Americans made in centuries past. Then with my crooked knife I began shaping the stick. I'd first spotted the shelter a couple of weeks earlier hidden in a mott. Arched into a wooden igloo of sorts the poles were covered with smaller branches and leaves. A crude shelter made in an attempt to remain concealed, it was no doubt the work of what I'd taken to calling *long-distance travelers*— those who trekked north from places far away. Granjeno is most likely the keystone hardwood on the western half of the Sand Sheet. Its arrival in the desert probably occurred as birds lit on herbaceous shrubs and dropped seeds onto the ground. Some have suggested that humans might have helped spread granjeno across the sands because its small, bright orange berries were a valuable food source. But I don't find that second scenario as plausible since birds, deer, javelina, coyotes, and raccoons eat granjeno berries. Because granjeno grows in clumps its fallen leaves form mats that fertilize the sand, thus creating a foothold for other hardwoods. Brasil, colima, and lotebush produce edible berries as well and thus live synergistically within the granjeno motts. In another year or two the branch I'd collected from the discarded shelter would have rotted away. But not this stick,

I thought, and was reminded of the circle common to people in prehistoric times. Now the little branch would provide another step in completing that circle. From a rusted mill file that became a crooked knife to a weathered branch turned into a walking stick that might likewise have been made into the leg of a camp chair or perhaps the trigger for a deadfall or maybe a rooting dowel or a dozen other things that can be fashioned and have been made, including a wickiup. Although not by me as much as by a thousand generations past that lived without machines rolling or flying and that never heard noise beyond a clap of thunder, and saw no other light but that derived via the sun and stars or the embers of a fire or an occasional bolt of lightning. For them every step in the making of things brought the circle nearer to closure. For like the file that became a knife, the maker took a rock and knapped out a blade. Or perhaps it was a mound of clay scraped from a barranca then mixed with sand and water and molded into a bowl or cup. Maybe it was grass gathered from a meadow and woven into a basket, or fibers pounded from the swollen leaves of an agave then rolled into cordage. Every man and woman knew the circle and could complete it on his or her own. But it is not like that today for the circle no longer has meaning. There is no connection other than an ethereal idea of how things might have come about. Nothing is held in value but that it amortizes over time and therefore its worth has no importance leading backward but only as it decays into the future. For there is no rock to knap into an atlatl or arrow point or reeds foraged to weave a basket or stave shaped to make a bow. The circle has become a line of dashes and dots occurring separately and without connection. Our very being is tied not to the circle we complete but to the diversions we might find to thwart our boredom. Even our truths are but reflections of our own self-absorption.

I've made dozens of crooked knives. Their origin is shrouded within the deep snows of the Canadian northeast. Among the Algonquin tribes the crooked knife found its beginnings. Designed to be pulled and not pushed as with most woodcarving knives; the upward sweeping motion is controlled by the finely innervated bicep. A razor sharp chisel bevel allows the woodcarver to peel off thin layers of wood using smooth and graceful strokes. As such the crooked knife allows the woodcarver to caress the wood instead of holding it at bay. Some people call the crooked knife a one-handed drawknife, but it accomplishes far more carving tasks than any drawknife.

Because the granjeno branch was dry it offered a few challenges. Woodcarvers prefer green wood because dry wood, though less likely to check, fatigues the hand and can quickly dull a knife. When making spoons or bowls I often immerse dry wood in water for several days in order to engorge the fibers and save my wrist and fingers the anguish of battling desiccated wood. So I worked slowly on the branch sweeping the crooked knife upward and shaving off paper-thin slices until I beheld a walking stick. I now had the tool I needed to push aside branches growing across the trail or prod tufts of grass to see if anything rattled. The walking stick also provided support when I bent to avoid low-hanging limbs. In years past I would never have considered using a walking stick. But by the time I arrived at the Sand Sheet things had changed. Now I never took to the trails without a stick in my hands. Besides, the stick allowed me to move through motts silently. While others might use a machete to whack away branches blocking the trail, I used my stick to quietly move them aside. I walked ever mindful of my surroundings, looking out across the flats of dried grass and withered shrubs and then around me within the motts that appeared like islands amid rippling waves of straw. As a boy we'd sometimes use long sticks to dispatch rattlesnakes hidden in shrubs or under clumps of cactus. On one occasion a friend and I were trapped between a stand of nopal and a six-foot rattler. There was no place to step back so my friend reached to his side and snapped a dry mesquite branch from a dead tree. Slowly he moved the branch over the rattler's head and then just as it seemed the snake was about to strike he thrust the stick downward to pin it to the ground. A quick jab and twist and the snake was usually no longer a threat. Except this time the stick snapped in two at the very moment of impact and my friend lurched forward. Somehow he was able to keep from falling for he was well motivated and even though the snake sent its head outward like a missile no contact was made. From that point on an amendment, perhaps in the form of a mental footnote, was attached: Never use dry limbs.

6

After living a year at our new home my life settled into a routine. I'd kept a blog called *Woods Roamer* recounting my days in the woods as well as posts on the acquisition of primitive skills and the native plants of the region. I wrote about making knives and selfbows and about woodcarving and, of course, saving nature. I received hundreds of e-mails from people saying they envied my lifestyle, living in a secluded cabin in the woods. But I think a man or woman who truly wants that life will find a way to make it so. For that reason, I believe most people who commented about living in isolation were speaking metaphorically even if they didn't realize it. For them a cabin in the woods represents fleeing from a world that has become too stressful, too crowded, and oftentimes too depressing. It's not that they'd be happy living in such solitude but that they perceive the concept as a way out. Perhaps that is why many of those who sent me letters seemed eager for the world to end. In the 1970s it was the survivalist movement that eventually morphed into what people today call "preppers." Even so, end of the world scenarios hark back over two thousand years and have usually focused on an ending that leads to a new beginning. I believe that is at the heart of what people want when they speak of going off and living in a cabin in the woods or when they talk about a universal collapse of one sort or another. They want an end to the way their lives have been scripted. They think such an end will lead, in convoluted and oftentimes unrealistic ways, to a new and better life. Living in the woods, however, is not a panacea and it suits only a small number of people. Invariably visitors found the place too remote, too quiet, too scary, or too difficult to maintain. Some complained that the woods surrounding our cabin was too closed in and suggested we clear the lower branches so we could see through the underbrush in case someone might try to sneak up. I'd let them talk, realizing they brought their city attitudes with them and it would be impossible for them to understand life as I saw it. It seemed that for many there were monsters in every shadowy spot and their nighttime anxieties were almost palpable. I offered nothing more than a smile when they insisted on locking their car or pickup doors when they parked in the driveway. While I thought the tranquility would bring them solace I was surprised to see the opposite reaction. Many visitors seemed tense from the moment they arrived. Gone was the undercurrent of continuous traffic or the sounds of sirens and loud music. Instead, there was only nature's silence, a mixture of birds' calls and whistles or the gentle swoosh of breezes blowing through the trees. The sounds, though natural, were unnatural to many of our visitors, and we became used to them saying, "Aren't you lonely?"

Living in the woods requires a learning curve, and one must acquire basic skills. It wasn't as if I could always call in a plumber or electrician or carpenter for a quick fix. Likewise, water wells with submersible pumps are temperamental contraptions. Over the years, however, I'd developed a mindset of self-reliance and minimalism, an attitude that served me well in the woods. But as weeks turned into months and the first year became a memory I was reminded that plans often fall short and that living remotely requires constant alertness and carefulness. One afternoon when I went to feed my six blue heelers one was missing. Her name was Chucha and she'd been part of Pita's litter. I'd given Chucha and her sister Maggie to a relative, but when my relative was mowing her lawn Chucha somehow got run over and lost one of her front legs. After I moved to the cabin I reclaimed Chucha and Maggie. Every morning I'd treat them to dog biscuits and in the late afternoon I'd feed them.

Afterward I'd take them on three-mile walks down the road leading back toward San Isidro. Chucha might have only had three legs but she was not about to be held back. Like all blue heelers she loved chasing cows. Tololo Lopez owned a few cows on a nearby piece of property, and my heelers thought of those cows as their own. On nearly every walk they'd take a few minutes to herd the cows safely near the water trough. Then they'd stand guarding them and looking at me as if to say, Aren't we a fine bunch? But on that afternoon when Chucha was nowhere to be found I realized the dogs had been acting strangely since morning. Chucha had shown up to get her morning treat but hadn't eaten the biscuit I'd given her. The two oldest, Dingo and his sister Chula, had huddled outside next to our bedroom window facing the back porch. The other dogs, Pita, Maggie, and Oy, kept close to the front door. As Norma walked along the porch calling out for Chucha she heard a high-pitched buzzing coming from underneath the house. She recognized the sound immediately and called for me. One shot from a .410 shotgun ended the rattlesnake's life, but when I peered into the darkness beneath the house I saw Chucha's lifeless body. It was already too dark to crawl underneath, and during the night I walked out on the front porch several times thinking about Chucha lying still but a few feet away. The hours slowed as both Norma and I waited for morning. Shortly after sunrise I recovered our little blue heeler with the big black spot on her face and then we held a private burial not far from the house. With only three legs Chucha was unable to outmaneuver the big rattler, and when it bit her on the face the venom quickly entered the bloodstream.

There were other rattlesnakes and even coral snakes that came into the yard. I was forced to shoot several of them. A few indigos called our place home as well. Occasionally one would slither up to a watering station for a drink. The green jays and brown thrashers pestered the indigos as would the mockingbirds. But the bulky black snakes paid them no attention. At night around my shop I'd see Texas blind snakes drilling into the sand and occasionally a scarlet kingsnake slithered across the concrete blocks forming the floor and took shelter behind an old bookcase. But the snake that intrigued me the most was about thirty inches long with a triangular head, its scales covered with black and coppery splotches. On warm nights it would coil up in a water-filled dish under one of the faucets. If I bothered it too much with my flashlight it would crawl off to one of the other dripping faucets and again curl up and not move. We had leopard frogs in the yard and I'm sure the northern cat-eyed snake was there looking for a meal. I suspect it spent most of its time in the thick foliage around our gray water pond where frogs abounded, but it occasionally slipped into the yard to see what pickings were available under the ever-dripping faucets. The northern cat-eyed snake is moderately venomous, and although I would approach it closely it seemed docile and not aggressive.

An acquaintance who loves nature as I do asked why I shot rattlesnakes instead of catching them and releasing them away from the house. I told him, "I shoot them only when they're in my yard," and then added, "If you discovered a black widow spider in your daughter's closet would you catch it and let it out on your patio?" He said, "No, but if I found a rattlesnake in my backyard I'd capture it and release it in the woods." And to that I replied, "Into my backyard?" But I always left rattlesnakes alone when I encountered them away from the house. Some weeks later I told my longtime friend Benito Treviño what the man said to me regarding rattlesnakes. Benito was outraged. "But he wasn't there!" he said. Then he said his maltipoo puppy had been killed by a rattlesnake at their home outside the town of Rio Grande City. "I found the snake near the house," Benito said. "With the help of a man who lives with us we caught it and transferred it into the deeper woods about a mile away. It was a big snake and I noticed it had peculiar markings that made it distinct. We had the cutest little puppy we'd let out of the house now and then. I guess I never figured anything might go wrong. But a few days later I spotted the same big rattlesnake coiled up next to our house and then I saw our puppy. She was in bad shape. I killed the snake and grabbed the puppy and drove as fast as I could to the vet's office. But it was too late. She died in my arms. The vet said she'd been bit three times."

I was raised among rattlesnakes and have encountered them thousands of times in my life. But in all those years I'd never experienced a loss brought by a rattler. Losing Chucha was difficult and I realize the snake was only defending itself. Nonetheless, I could not afford having such dangerous creatures so close to the clan.

Chucha's father Dingo and her aunt Chula were both fifteen years old. Chula was born deaf and many found it amazing she'd lived so long. Take care of your dogs and they'll have good lives, I've always said. But a few months after Chucha's death our beloved Chula said goodbye and then several months after that Dingo left us as well. Brother and sister, they helped raise my two youngest sons. When Dingo was in his prime he always accompanied me into the woods. When we'd take a break I'd grab a granola bar for me and a dog biscuit for Dingo. But he wouldn't touch the biscuit and instead would stare at me as if to say, "We're partners, old boy. I get what you get." So I'd break the granola bar in two and give one-half to Dingo. Within an hour of his death I felt compelled to write and so I started a new page in my blog *Woods Roamer*. Dingo was the king, I said. He was the greatest blue heeler I've ever known. He turned sixteen years old this past summer. He was blind and deaf now from old age. His teeth were nearly all gone. But in his youth he was fierce and no one messed with us under any circumstance. Even in his old age he always went walking with me, keeping close, following my scent, and though he'd developed arthritis and sometimes had a limp he kept going. Dingo never complained. He was given medicine to help his joints and was fed special food to ease his chewing but he was always eager to go woods roaming. If the wind changed or if I happened to amble off the path, then Dingo would sometimes get lost and I had to walk back and find him and make sure he stayed close. I had thought about getting him a leash and I'm sure he wouldn't have minded but somehow I just couldn't do that to him. He was too regal to be walked with any sort of cord around his neck. Besides, we live in the woods and only city dogs get paraded around that way. Dingo was free to take the path as he wanted. Lately, Dingo's eyesight was getting really bad. I think he was nearly completely blind suffering from cataracts and perhaps he could only make out vague shapes and colors. When I'd call him he couldn't locate me and I'd have to move around to let him know where I was standing. Besides his poor eyesight, his hearing had become nearly nonexistent and yet amazingly he could hear certain types of sounds. You see, the US Navy has an airbase about 150 miles to the northeast and they sometimes train in dogfighting overhead. They figure since no one lives out here but a grizzled old hermit named Longoria it doesn't matter if they chase each other at 20,000 feet. I don't pay them much attention since it sounds like thunder high overhead. Besides, they only dogfight about once or twice a month, and I figure I can put up with those rumbling jet engines for a few minutes as part of my contribution toward national defense. But the dogfights drove Dingo crazy. He'd start yelping and crying and moaning as if he were about to get attacked. Maybe he thought it was wolves howling in the distance. So when the jet fighters chased each other overhead Dingo would start pleading for mercy. It never failed. Bring the jets and Dingo started wailing. Now there's a little road about seventy-five yards beyond a thicket in front of my cabin. It's the two ruts I take to get to the first locked gate on the way out to the world beyond. Dingo liked to sit at the end of our driveway keeping guard and looking out on that little road. Granted he couldn't see or hear anymore but nonetheless he'd station himself out there just in case—just in case of what I'm not sure but anyway, just in case. I think every dog yearns to chase cars, and Dingo spent his time out there waiting for the car that never drove past since nothing comes by except an occasional wandering coyote, a trail of leafcutter ants, free-ranging dung beetles, or manic roadrunners. But Dingo was a positive thinker and he was out there just in case. As it turned out I didn't take Dingo walking yesterday because I was tired. I was up before daybreak and at sundown I was still working and after a shower and supper I drifted off. At sunrise I got up and made coffee with my usual oatmeal and blueberries and homemade date/cranberry bread with peanut butter. Gave the doggies their treats and noticed Dingo out at the edge of the driveway asleep. Sent Maggie out there to wake him up. It was a nippy morning and Dingo was awake in an instant and trotted back for his cookie. That's my last impression of my beloved Dingo. You see he finally did get his chance to chase a vehicle. But he was blind and it ended badly. I buried Dingo at the edge of the driveway looking out on the two rut road that leads to the first gate. I think Dingo would like that. Just as I was packing down the dirt around the grave a couple of US Navy fighter jets flew overhead at about 10,000 feet. I could've sworn one of the jets dipped its wing and dammit but I think I actually saw the pilot bring his hand up and offer a salute. Yep, I'm pretty sure of it. Dingo couldn't cry back like before but I'm doing a little bit of that now for him—if you don't mind.

※

In the span of a little more than a year we were down to three blue heelers. Pita, the mama, was nine, and her two surviving pups, Oy and Maggie, were seven years old. But every day in the woods brings new adventures, some good and others full of problems. Several months later I was taking Oy and Maggie walking and they heard a noise and ran headlong into the brush. Then came a high-pitched squealing and my first thought was that they'd run into a group of wild hogs. Maggie bolted out of the brush, but Oy didn't follow. Instead, I heard an intense fight. Armed with an ancient .22 long-rifle revolver stoked with snake shot I could do nothing but stand on the road listening to the battle. At last Oy limped out. Blood was spurting from his right flank, and if that wasn't enough two javelina boars were in pursuit. The javelinas hadn't noticed me but instead seemed intent on Oy, who marked his trail with a line of blood squirting onto the ground. I pulled my revolver and shot into the air as Oy stood by my side bleeding profusely. Only then did the javelinas see me, and they turned and ran off. I stooped to examine Oy who was panting hard as a pool of blood formed on the sand around him. I called Norma and she arrived within a few minutes as I pressed my bandana into the wound trying to stop the bleeding. Norma brought a towel and applied pressure on the wound. I thought Oy had bought it but finally the bleeding stopped and we lifted him into the bed of the pickup and drove him back to the house. In a month he was back to normal, but about six months afterward Oy met up with a rattlesnake at the end of our driveway. He had a habit of pouncing on kangaroo rats in the grass, but this time something else lay coiled and in no mood to play. We never found the snake but I was in my workshop when I heard a horrific yelp. I found Oy lying down with the strangest look on his face. He was already going into shock although I had yet to realize he'd been snake bit. We examined him and it was then that I found the fang marks on his left leg. Fortunately, only one fang pierced his skin and the other glanced off. But Oy was still in critical condition. His heart began racing and his breathing grew labored. We were too far from any veterinarian, assuming we could even find one open by the time we got to the city. But we had antibiotic on hand as well as a steroid for our dogs in case they got snake bit, so we administered both and then tried our best to comfort Oy. His eyes became severely bloodshot and his leg began swelling. We kept encouraging him to drink water but those first few hours weren't easy. Over the next several days Oy began to come around. We kept hand feeding him and at last his condition started improving, even though it was months before he regained any weight. Six months later Oy was still limping and I feared he'd never run again. It took a year for him to finally get back to normal, but I don't think he ever pounced blindly into the grass after that.

You'll pet your dogs and love your dogs and talk to them as if they were human. But then maybe they are or at least there's something so special it transcends our understanding of what it means to be human. When they cast their eyes on you they look into your very soul. Their lives are far too short and when they're gone there remains an empty place in your heart. When I was a boy there was Sally and Freda. When the older boys, Nomar and Jason, were little there was Girl, a shepherd mix, and Husky, a Norwegian elkhound, then Frisky, a border collie, and Coco, a beagle. Dingo, Chula, and Chucha were, like the others, as good as they get. Family members, loyal, loving, and fierce protectors. You never forget your dogs.

7

I'd sometimes rise at four or five in the morning missing my children. Walk out on the front porch and stand watching clouds sweeping across the sky northwestward. Other times endless stars dotted the heavens, and the trees around me appeared as nothing more than a greater blackness in a world gone silent. I'd walk the little road for a mile or two. Flashlight and canteen, a knife and walking stick, and my musette bag. Occasionally, I took a coffee pot and bag of coffee, some creamer and a spoon. The predawn is marked by subtle hues of pink flaring overhead from the Gulf Coast and the sounds of birds greeting an awakening sun. At dawn I'd spot a deer or two ambling from their feeding locales to their bedding places. Or maybe I'd see a coyote stalking across a field in search of one more mouse. If it were not for the woods, I think I might not survive. I'd gather a few sticks and make a small fire then set my coffee pot atop the coals and sit on my folding stool allowing the stillness to bring me peace and make me well.

I recall those who came to nature desperate for healing and yet anxious to get it over with all at the same time. They chanted and danced and consumed copious amount of drugs and spent their nights with devils and demons and when nothing happened they packed up and filtered off disenchanted and lost. They spoke of living off the land but starved because they did not take the time to learn any skills. They wanted their suffering cured but were impatient seeking a fix as if brought by a magic wand. They draped themselves with beads and charms and upped their drugs and still the answers eluded them. And then they became angry. Because what they had thought would heal proved a failure and now there seemed no way out of the increasing madness surrounding them. For some the answer became one of compulsive hedonism where the hallucinogen was replaced by lavish consumption. Where they had once looked for quality in their lives they now searched for quantity, and the difference between the two became increasingly blurred. Others walked away with a broken spirit. Nothing made sense anymore. All was hopeless. And yet, for some it started with a journey into nature. So where did it all go wrong? How did nature fail to bring them what they so desperately sought? Perhaps it was because they did not understand the stillness.

I drank my coffee then smothered the little campfire and walked on, following a trail that weaved between clumps of small trees and woody shrubs. Deeper and deeper into the woods as the rising sun peaked over thunderheads far to the east. At a bend in the trail I found a cow's skull, and since I'd left my folding stool back with my coffee pot the skull became an impromptu place to sit and watch the woods around me. As the stillness deepened I became increasingly aware of my surroundings. This was not introverted meditation but rather an outward awakening. I did not close my eyes and seek darkness but instead allowed light, sound, and smell to escalate as I grew more aware of the woods' every nuance. Soon I was but a plant growing from the cow's skull and the morning sun but a flickering light piercing through the dense brush. The aromas of lippia and salvia intermingled like spices mixed in a recipe. A mouse followed a tiny trail winding between clumps of grass and I felt each blade brushing the rodent's fur. Crickets croaked in concert with a mockingbird's song, and a brown thrasher danced along the branches of a mesquite tree. The stillness is the acceptance and celebration of what is real, and as such it becomes the one thing more than any other to be cherished. All else is but speculation or myth confounded by time and vanity. Our imaginings too often reveal nothing more than a shameless materialistic quest, an abandoning of truth.

A covey of bobwhite quail walked toward me scratching at the ground and pecking things here and there, a continuous communiqué with soft and barely audible chirps and whisperings. Closer and closer until they were at my feet and I examined each feather's shade of color. A Harris hawk called out from the deeper woods and the quail scurried into a stand of granjeno. I heard an animal rooting at the ground a few yards away and smelled the sweet odor of javelina.

Before sunrise I'd heard the odd rumble of a passenger jet and looked up to see a listless flickering light crossing the heavens. From a pond south of me came the yodels of coyotes. Walking the road in the darkness I glanced across the horizon and saw scattered within an esoteric randomness a half dozen pulsating red lights where radio towers marked lonely places on this earth. At a bend in the road I heard a faint sound and so I turned on my flashlight and spied a kangaroo rat hopping along a fence line, stopping at each post before proceeding onward. But when the day arrived in full and the heat of late summer, *la canícula* the locals call it, settled around me I nudged the cow's skull back to the spot it had occupied and walked away.

One night I hiked a couple of miles to the second gate, and as I walked back to the house I noticed two orange-yellow eyes reflecting back at me from my flashlight. I stepped closer but the eyes did not move. Closer still until we were but a few yards apart. The orange-yellow eyes seemed as curious of me as I was of them. A bush partially concealed the eyes so I moved to my left, inching closer until I was but twenty feet away. The eyes turned and my flashlight illuminated something graceful and muscular. The cougar stood a moment then scampered into the brush and disappeared. That too was part of the stillness, for it is nature emerging. A breeze flutters through the trees or whistles across a rocky outcrop, or perhaps an eagle shrieks from a precipice or a dove coos from the branches of a palo blanco. Nature arrives on its own terms. You cannot push nature though many have tried. Those who came with visions of salvation did what others have done for generations: They forced nature to serve them without considering the extent of their demands. That was their mistake; nature does not act favorably when coerced nor does it respond well to domination. Nature governs in accordance to its own laws. It cannot be tethered. Those who came to find truth and turned away in bitter disappointment crowded out the stillness as others have done before and afterward.

8

Autumn arrived and then winter. I used to enjoy the hunt in fall. But deer and quail season in Texas has become less about hunting and more about business. From early November through the end of February I'd hear gun shots. Some far and others close. It seems the locals had mixed views about those who came to hunt deer and quail from cities to the south and north. I heard opinions ranging from "I can't stand them" to "they're good for business." In October they'd begin hauling in their shooting towers and travel trailers. On one of my weekly visits to the post office in San Isidro I ran into a fellow I knew and the subject of out-of-town hunters came up. "You can always spot them," he said. "In their camouflage uniforms and snake boots and talking too loud." Just then a couple of 18-wheeler trucks drove by going about 65 on a road marked 55 mph. "Damn trucks," the man said. "They drive down 1017 all hours of the day and night and then the hunters show up and the only thing for me to do is head out to my family's land and stay there until I calm down." We talked a bit more on other things but I guess he was still thinking about city hunters because as I was getting ready to leave he said, "I guess we look as odd to them when we're in the city as they look to us out here."

I met a man who lived in the Rio Grande Valley to the south. He owned an eleven-pound sniper rifle with a fiberglass stock and 6.5 × 20 power scope. "What caliber is it?" I asked. He looked at me proudly and said, "It's a 7 mm STW and I'm here to tell you I can reach down a sendero with this one." He bragged about sitting in his shooting tower and "nailing deer eight hundred yards away." On another occasion I heard someone calling out from the woods as I took my late afternoon hike down the dirt road leading away from the house. So I stepped into the brush and walked about thirty yards and saw a man sitting in a small shooting tower. "I need to pee," he said. I nodded and replied, "Okay." But then he said, "No, you don't understand. There's a bunch of hogs in that brush right there." He pointed at a nearby clump of thorny shrubs and said, "I don't want to get down 'cause of the pigs." I'm not sure whether I wanted to laugh or just felt sorry for the man. He was as out of place as are many others who journey to the woods to hunt. Like the man from Houston I met years ago who came to hunt in Deep South Texas. He told me he was carrying a three-hundred mag because "that really puts 'em down." I asked if he'd ever shot a deer, and he shook his head then admitted he'd not even shot the rifle he was carrying. "But it'll put 'em down for sure," he repeated. My buddy Mario Hernandez climbed into the shooting tower with the dude from Houston and I walked off. But later that evening Mario said a nice buck had crossed the sendero about a hundred yards from the tower. "That's a good one," he whispered to the man sitting next to him. So the man aimed his magnum rifle and then to Mario's astonishment started ejecting all the cartridges from the magazine. In rapid succession the shells clanked onto the floor of the tower. Then the man, out of breath and barely able to talk said, "Did I get him?"

"Why don't you just pee from the blind?" I asked the man who'd called for help.

"Because there's not enough room for me to stand up properly," he said.

"Okay, do you want me to go in there and chase away the pigs?"

"Would you mind?" he asked.

So I walked into the brush and if there were any hogs they had long since moved away. I didn't find any fresh tracks either but the man swore he'd seen a giant boar. "It's a monster," he said.

"Why didn't you shoot it?" I asked. But he'd just finished peeing and was eager to get back into his shooting tower.

"Park your truck next to the blind so you can hop into the bed directly from the ladder," I said.

"That's a good idea," the man answered. And though his truck was no more than about twenty feet from the tower he moved it next to the ladder exactly as I'd recommended and seemed quite content with the solution.

Another young fellow who appeared incapable of walking anywhere hunted hogs on a nearby property. He'd arrive on the weekends driving a pickup with the muffler removed. I'm not sure what that was all about but the Sand Sheet cannot tolerate vehicular traffic, and I was never too keen about having to listen to that moaning rumbling noise in the distance. Sound travels in quiet places. I could always tell where this fellow and the other hunters had driven as the ruts remained etched deeply into the sand. But on one occasion I met a man who hunted with a compound bow. He too was in camouflage but he sat on a wooden stool pressed up against a chapote tree watching a narrow trail that crossed in front of him. When I met him he was walking back to his truck parked alongside a narrow caliche road. A neighbor had given him permission to hunt that day. He wore a pair of heavy leather moccasins and carried his arrows in a plastic quiver attached to his bow. We both said hello and then he asked if I was hunting.

"No," I said.

"You just like walking?" he asked.

I nodded and smiled and asked, "Have you seen anything?"

"A couple of fat does," he said grinning. Then he shrugged and said, "To tell you the truth, I just like sitting under a tree and enjoying the quiet."

9

We owned no television but a forty-year-old copper cable provided sporadic Internet service. The cable had deteriorated to the point that in some places it was flaking into dust. When technicians from the telephone company arrived they always said the same thing: "You need a new cable run out here." But the telephone company wasn't willing (at least not in the beginning) to install a new cable. So my Internet connection worked some of the time but didn't work most of the time. I'd kept up my blog, *Woods Roamer*. Blogs are curious things. They fall into niches. I was surprised to learn that many people who were interested in making knives and bows showed no interest in nature or in saving the environment. The most unconcerned seemed to be the survivalists or preppers who were fixated on stashing gear for the big collapse. One day a reader sent me a link to a survivalist website where the members were discussing my blog. One fellow said he had stopped reading *Woods Roamer* when I began writing about saving nature. He said he made his money welding gas pipelines and working on fracking rigs and wasn't about to read anything that questioned the safety of fracking or the environmental damage done by pipelines. Besides, he needed his money to gather things for the collapse. I couldn't help wondering if he realized that his work was contributing to the environmental damage that might precipitate the very sort of catastrophe he feared. Regardless, if I wrote about a new bushcraft knife I'd forged I'd have a thousand readers. Talk about the coming water crisis or about the evils of fracking or rampant population growth and the readers were fewer though more passionate about the subject and much more willing to comment. *Woods Roamer* was about self-sufficiency, but the most popular "bushcraft" blogs were all about product reviews and what to purchase next. It seems a man who wears ten dollar blue jeans, twelve dollar shirts, cuts his own hair, makes his own knives and self-bows, grows his own food, and lives a secluded lifestyle doesn't fit well into any capitalistic cubbyhole. Besides, making things in my little shop at night was a form of therapy and it wasn't all that important to share that with the world. Taking a 5160 leaf-spring or piece of 01 steel, or an anacua or elm stave, and turning those things into knives or bows gave me peace. They took me into that world Heidegger termed *Dasein*, or what Dr. Speeg called "flow." Immersion into a project made other concerns disappear as well. When I was building a new bow or forging a knife I forgot, at least for those moments, how much I missed my children. I'd forget about those I loved now gone. I'd not dwell on all the years passed. A pleasant distraction bolstered by night birds whistling and hooting and by stars and moonlight and the stillness.

I kept writing about knives and bows and now and then about native plants and the environment, but more than anything the blog was simply a way of saying, *Hello world, I'm still alive.* My scheduled three-month lab test results remained good and I kept to my walks and diet. Still, there were things going on around me that were disturbing. I started getting reports that the oil and gas companies were busy fracking parts of Texas into oblivion. I'd see trucks carrying fracking fluids speeding along FM 1017. I'd pass by deep well injection facilities. I heard about earthquakes in North Texas and Oklahoma caused by both processes. I'd read about the millions of gallons of underground water consumed by each drilling operation. Toxins and carcinogens like xylene, toluene, and benzene pumped without remorse into once-pure aquifers. Then on one of my visits to San Antonio someone mentioned that well workers in the Eagle Ford

Shale Region were coming down with severe liver disease. The doctors at the hospital had already performed several liver transplants. There were also reports of miscarriages and birth defects attributed to air and groundwater saturated with toxins. Perhaps the devil does indeed have angels. Whether disguised in suits or driving trucks filled with poisons, they bear no shame while delivering us all into evil.

It seems the Sand Sheet is blessed and cursed all at the same time. Left to its own druthers and the sheet continues its long journey back toward diversity. Add humans to the mix and things too often slip into disarray. Despite everything, however, the Sand Sheet always seemed willing to give of itself. And yes, there are even millionaires on the Sand Sheet. They live idle lives searching for things to do. Their land is pockmarked with oil and gas wells and crisscrossed with roads and pipelines and terminals and buildings, and they drive around like strangers with little power other than what the oil and gas companies allow. The royalty checks arrive and they want for nothing other than perhaps the satisfaction of having made something for themselves aside from what has been given to them at the expense of destroying all that surrounds them.

Even so there were respites on the Sand Sheet, times when we were able to put that world of avarice and destruction at a distance. Small things that to some wouldn't mean much though for us they defined the beauty of life. There was Buddy, for example.

Every morning we'd put grain in three clay pans that dangled from agave cordage tied to mesquite limbs in our front yard. We'd also scatter grain on the ground for the bobwhite quail. The birds would wait in the brush surrounding the house and as soon as we were back on the porch they'd swoop or run into the yard. But a young male bobwhite always came out to greet us. He'd approach gingerly and we'd set a handful of grain on the ground just for him. Then we'd squat by his side as he ate. We named him Buddy. After a while I stopped placing the grain on the ground but instead kept it in the palm of my hand, and Buddy would peck each seed until it was all gone. In case you're wondering, the pecks are forceful and take a little getting used to. One morning I made a video of Buddy eating his breakfast. Buddy was as wild as any other bobwhite quail, but like the others he'd learned we weren't dangerous. Buddy, however, took it a step further. The word *harmony* comes to mind when I recall the two of us walking side by side in the front yard. I called my friend Ruth Hoyt, who teaches classes in wildlife photography, and told her about Buddy. Then I sent Ruth a still from the video. Of course, I figured when quail season arrived I'd probably never see Buddy again. Then the season opened and as we'd suspected Buddy disappeared. We'd sit on the front porch thinking about Buddy and about the times we'd feed him. For a while all the quail vanished and even though we'd hear their calls in the distance they no longer came into our yard. We kept placing grain on the ground and the doves and songbirds always left the plate clean, so to speak. Then one day in mid-December I looked out the front window and saw a covey of quail munching at the grain we'd tossed out earlier. I watched them for a few minutes and then decided to get more grain and sprinkle it out for the birds. I grabbed a handful of sorghum and walked into the yard fully expecting the birds to flush. The covey darted into the surrounding brush all except for one male that stayed in the area where we'd placed grain earlier. *It can't be*, I thought. Then I squatted and said, "Come here, Buddy," and held out my hand. The emotions run strong even now, but the little male approached me and I knew it was Buddy. He'd grown a bit but was just as friendly as before. He pecked at the grain with timid tweets and chirps. "Hello, Buddy," I said. "I'm so happy to see you."

Not long afterward Buddy showed up with Mrs. Buddy and his family of seven chicks. The little bobwhites had no idea tradition demanded they be afraid of all people. But the little ones approached me as eagerly as their daddy, and even Mrs. Buddy became friendly when she saw the others standing around me. Twice a day we'd walk into the front yard and Buddy and family would run to us like chickens. With the dense brush surrounding them they'd found a haven from predators and people. In time other quail saw we posed no danger and within a few months we had dozens of bobwhites following us around the yard. Other birds arrived to greet us as well. Cardinals, thrashers, Inca doves, titmice, pyrrhuloxias, mourning doves, ghost doves, and woodpeckers; the extended family grew weekly. A screech owl we named Henry took up residence in an owl box I set at the edge of the yard. Every morning when I'd walk out of the house calling, "Okay, everybody, come on everybody," Henry would peek out of his house, eyes squinted as if to say, *Can't you see I'm trying to sleep here?*

10

One night I heard a horrific ruckus on the front porch. My dogs went wild. I opened the door and standing not more than four feet from me was a dark-complexioned man with tattoos on his arms and neck and even on his face. The realization was instant. I'd made a horrible mistake by opening the door without a firearm in my hand. On his arms and on either side of his neck was tattooed MS13. He was a member of the Mara Salvatrucha, a notorious El Salvadorian gang considered by many one of the most violent criminal organizations in North America. Had it not been for my dogs I don't know what might have happened. I reached back inside the door and Norma handed me a pistol and I held it on the man. He scowled at me and said, "Your dog bit me." Thank you my dear Pita, I thought. "My dogs are vaccinated," I said and then asked, "How many more are in the woods?"

I'd been warned by the Border Patrol to be on the lookout for violent gang members crossing into the United States through Starr County. The Mexican cartels were working with the gangs to bring narcotics and people into the country. I'd seen tracks in the woods where convoys of smugglers and those who'd entered the United States in violation of immigration laws had walked through the area. A few weeks before, as I was shooting one of my bows in front of the cabin, I'd spotted two objects just beyond the wall of trees about fifty yards to the east. One of my arrows missed the target and when I went to retrieve it I saw the objects were two people lying down and trying to hide. I jogged back to the cabin and grabbed my shotgun and when I returned to the woods I saw two women running away.

"Stop," I yelled.

The women turned around to face me.

"Come over here," I said.

They too were from El Salvador and one said she was fifteen and the other said she was seventeen.

"We're lost," the older one said.

As usual their smuggler had slipped away while they were sleeping and they'd spent part of the previous night and all that day walking along a fence line hoping to come across a road.

"You need to let us stay here with you," the older girl said.

"I can't do that," I told her.

Norma gave them water and a couple of sandwiches and then I told them they wouldn't survive if they tried to cross the desert.

"Then let us stay with you," the older one insisted.

"I'm sorry but we can't do that," I repeated.

The older girl looked threatening and I was surprised by her attitude. She kept insisting and I kept refusing. At last I called the Border Patrol on my cell phone and the dispatcher said there were agents in the area. One of the agents called me and asked me to bring the girls to the first gate. I wanted him to make that decision because I wasn't about to be accused of transporting aliens. The younger girl didn't say anything but I could tell she was scared. But the older girl looked dangerous. I told them to get into the bed of my pickup and with Norma driving and me sitting in the back with the girls we headed for the first gate. When we reached the gate three miles away I saw the Border Patrol vehicle parked nearby. The girls climbed down from the bed and the older one kept looking at me like she'd slit my throat if given the chance. The younger girl had a sweater in her hands and she tossed it on the ground.

"Wait a second," I told her. "You're not going to litter around here."

This too I'd seen countless times. Clothes, empty canned goods, plastic water bottles, and trash bags in piles all strewn along trails leading all the way back to the Rio Grande sixty-five miles to the south. I'd found other things as well. On one occasion I discovered a Nuestra Señora de la Santa Muerte (Our Lady of the Holy Death) medallion lying on the ground. Drug smugglers often carried Santa Muerte medallions with them for protection.

I asked one of the Border Patrol agents how things were going.

"We're being overrun," he said. "We just found the one-hundredth dead body for the year a few hours ago." They'd also found a five-year-old boy wandering alone in the brush that morning.

As Norma and I drove back to the cabin I noticed something peculiar lying in the dirt just passed the second gate. I stopped and saw an identification card issued by the El Salvadorian government. I recognized the photo on the card immediately. It was the older girl except she wasn't seventeen years old but instead thirty. She'd thrown her ID away probably figuring that if she'd fooled the old man then maybe she could fool the Border Patrol. Minors are treated differently than adults.

The MS13 gang member was giving me a cold stare. "There're seven more in the brush," he said. It was dark and the dogs were barking fiercely towards the north. All except for Pita who was growling at the tattooed man and letting him know if he lurched forward she'd take his head off. "Don't move or I'll shoot," I said. Then I noticed I'd left my plastic canteen and a couple of granola bars on a metal rocking chair after I'd taken the dogs for their late afternoon walk. With my pistol held on the man I motioned for him to move back and sit down. He complied and I set the canteen and granola bars on the floor between us and said, "Now you've got food and water." It was my good canteen and I hated to give it to the man but I had no choice. I didn't want to open the door and ask Norma to make a sandwich. Unbeknownst to me she'd already grabbed a shotgun and was prepared to shoot anything that walked inside. Of course, I think I would've been an exception. Regardless, I figured if I showed the man an act of kindness he'd go away. He took the granola bars and removed the wrapper from one of them and gobbled it down in two fast bites. I kept glancing into the woods but I never saw the other seven men he said were waiting for him. He took the canteen and unscrewed the top and gulped down about half of the water.

"Leave now," I said. "If I see you or your friends I'll shoot."

The man was angry but he did as I'd instructed. He must've thought he could sneak up on the house and break in the door and then he and his buddies would rob whoever was inside. But he hadn't counted on our dogs, and they saved us that night.

Three days later driving to San Isidro I spotted something lying in a mud puddle. I stopped and sure enough it was my expensive canteen. The man had drunk the water and instead of holding onto it he'd simply discarded it in the road. I didn't even consider picking it up.

Our encounters with people were sporadic. They seemed to come in waves. I think when the Border Patrol covered one area the smugglers moved to another. A constant game of cat and mouse. So things would be peaceful for weeks at a time then people would start appearing. About four months after building the house, as I pulled weeds from around the gray water pond, I noticed the dogs looking intently behind me. I'd started carrying a firearm everywhere I went and always kept vigilant of my surroundings. I turned and looked to the west and saw three men approaching. The dogs ran toward them and the men stopped. I pulled my pistol from its holster and approached the men. One was carrying a backpack and I ordered him to leave it on the ground and then told the three to move away from the bag.

"Are you carrying any weapons?" I asked.

Two of the men, dark with long straight hair, looked in their early twenties and the other, equally dark complexioned but with short hair, was perhaps in his mid-thirties. He did all the talking. In fact, he talked too much. He said they'd been involved in a rollover a couple of hours before on FM 755.

"*La Migra* was chasing us," he said.

The conversations were always in Spanish but sometimes it was hard to understand the dialects when they were mixed with Indian words. Many of the people from Central America and Mexico are full-blooded indigenous natives. The majority come from the large cities but now and then I'd run into someone from the jungles or mountain regions.

"Everyone jumped from the truck and we all ran into the brush," the talking man said. He was wearing a blue baseball cap and I noticed he wore hand-stitched leather socks over his shoes secured with a draw cord. I'd known of people making sandals out of carpet remnants or inner tubes and wearing the sandals beneath their shoes to hide their tracks. Other times they'd grind the tread off their boots or shoes. But I'd never seen leather socks and his looked worn and black from use.

"We're walking to a stash house in San Isidro," the talking man said. "We just need some water to keep going."

So I told them to stay put and added if they tried to advance my dogs would attack. Norma's sister, Mary, and her husband Sal were visiting that morning and so Norma and Mary made three sandwiches and I filled a plastic jug with water. I walked back and found the three men still sitting on the ground but looking none the worse for wear after having been ejected from a speeding pickup.

"Was anybody hurt?" I asked.

The talking man shrugged and said, "We ran. We didn't look back."

"How many were in the pickup?" I asked.

But the talking man ignored my question and said, "Yeah, that stash house will be a good place to hole up for the night and we can get some rest."

"Okay," I replied. "There's your water and food. Now don't ever come back this way."

"Oh don't worry. We just need to get to that house in San Isidro to rest."

"Where are you from?" I asked looking specifically at the two younger men.

"El Salvador," one of them said.

"And you?" I asked the talking man.

"Oh, well, I'm from Mexico but I was on my way to Houston but now after the rollover I just want to go back and rest in San Isidro."

So the three walked away and I noticed the talking man took the lead as the two younger men followed. After they were out of sight I called the Border Patrol and asked if there'd been a chase and rollover along FM 755 earlier that day. "Yes," the dispatcher said. "Northeast of Santa Elena."

"Well, three of the people in the truck dropped by my place east of where the rollover occurred. One of them said they were heading to a stash house in San Isidro."

"Okay, I'll relay the message," the dispatcher said.

I finished pulling weeds around the gray water pond and

then a thought occurred. So I started following the men's tracks. Single file heading directly west into the brush edging clumps of nopal cactus, lotebush, and colima; the sun bearing down from directly overhead with a few clouds floating listlessly above. Wearing a long sleeve khaki shirt, faded blue jeans, a wide-brimmed felt hat, wedge-soled boots, and my .45 caliber revolver. My dogs walked alongside keeping an eye out and sniffing the breeze. But the tracks never turned south toward San Isidro. Instead, they hooked abruptly north heading back toward FM 755. I shook my head and mumbled, "Son of a bitch. He was the coyote." The talking man, the man from Mexico, the man with the plan. And the plan was to create a diversion; tell the gray hair pulling weeds part truth and part lie. Get everyone thinking about a stash house in San Isidro and instead slip back through the woods and keep going north into the desert where the smuggler probably had a couple or three carefully thought out routes, long traveled and well concealed. Places where neophyte Border Patrol fresh from the academy or even seasoned agents wedded to their vehicles, electronic sensors, and high-tech gear would never look. And they'd be gone. Hunkered down in a scooped-out foxhole during the day and vanished into the darkness come night.

Back at my cabin I didn't even bother calling the Border Patrol. By now the men would be well hidden within a thick granjeno mott shaded from the sun and sound asleep. Their stomachs were full and they had a fresh jug of water thanks to the man in worn blue jeans and tattered hat. Maybe *el coyote* had a cell phone and he'd contact someone to the south or farther north telling them there'd been a change in plans. They'd wait until darkness and a car would drive up FM 755 and at a designated location pull over and pick the men up. The smugglers know the land better than anyone. A *coyote* in his mid-thirties might have fifteen or even twenty years of experience. He learned from someone else who had twenty or thirty years of walking the brush and the Sand Sheet. The *coyote* is a master at reading sign and watching out for *la Migra* and talking his way out of critical situations when he runs into a homesteader or working cowboy. He can endure long hours of walking over soft sand and going without water for a day or more if need be. He'll forage off granjeno berries or *capul* or chew mesquite sap until he finds his way out of the brush. He'll stay off established trails always checking for sign and all the while listening for noises made by people nearby or vehicles in the distance. He'll crawl into a dense mott and not move for hours. If he has to go to the bathroom, he does it there. He'll cover himself with sand or dig into the ground as not to be detected by thermal imaging from the air. His woodsmanship is often far superior to any Border Patrol agent who goes in to find him. They might locate the people he was escorting but they'll not find him. He slips out of their grasp time and again, a phantom they speak of and even honor with a name. Respected and disliked all at the same time. Old-time agents will speak in almost reverent terms about *coyotes* they once pursued. They might find a track or two he set into the sand to throw them off but he disappears into the woods and vanishes. He knows every trick to convince natives from south of the Rio Grande, and Chinese from across the Pacific, and Ukrainians, Bosnians, and Irish and whoever else might come his way with several thousand dollars in his pocket that he is their ticket to the life they've dreamed about. They all know America. They've seen it thousands of times from William Holden and Nancy Olson to Tom Hanks and Meg Ryan. *La mesa ya está puesta*, my father once told me. The rest doesn't matter beyond that. The table is set in the great USA. It's just a matter of feasting. All they want is to eat and consume and be merry. Whatever might be happening in Honduras or El Salvador or Mexico or any other country will be somebody else's problem. It's never about making a country better but instead about getting to a country where everything has already been made better and then gobbling down as much as possible. India, Iran, Egypt, Pakistan, Serbia; I've met them from every corner of the globe. Politicians Right and Left think along those lines too, so it's just a game of getting what you can while you can. And yet I wonder if people ever stop to think about what would happen if everyone in their city decided all at once to come live in their house.

11

Our home was surrounded by wooded motts as if it were a tent enveloped by nature. Oftentimes I'd sit on the front porch looking out at the impenetrable clump of granjeno and brasil a few yards away and thinking about the animals and birds that made it their home. In the fall, winter, and spring I'd walk into the deeper woods and attach a hammock to suitable trees and then watch the sun fade in the west. Alone and yet not alone; a man's value rests not in what he acquires in life but in what he imparts to others. Sometimes I'd spend the night on the hammock. At dawn I'd eat a sandwich and brew a cup of coffee. I'd sit listening and thinking back. I recall decades past when I awoke on a hill overlooking the banks of the San Fernando River in the Mexican state of Tamaulipas. A crystal clear day and not a breath of air but instead a penetrating stillness, as if the world before me was a painting hanging from a wall. I was nine years old and it was the first time nature entered me, my first healing.

On a gray day in the second winter after building our new home I walked into the woods to the south and hung my hammock from two small Texas ebony trees. Clouds had pushed in from the north and the air smelled of rain. I stretched a lightweight tarp over the hammock and then sat watching the little fire I'd made. Nearby stood several tussocks of leather stem, knee high and leafless; the cinnamon-colored pedicles remain naked most of the year. But when rains come the slender stalks erupt with emerald green leaves that last only a few days. From green to yellow seemingly overnight, the leaves drop to the ground and are quickly reabsorbed into the earth. I had entered the transition zone separating the regions of clay soil and the Sand Sheet proper. The lines between these ecological areas are obscure. A geologist came to visit one day and asked me how I determined when I was on the Sand Sheet or on the transition zones. "The flora," I said. Indeed, the plants provide the key. I told a group at a meeting of Texas Master Naturalists that if anything is invasive where I live then it is the sand itself. For tens of thousands of years, the land underneath the sand was as ecologically rich as the land surrounding it. But when the sand arrived it buried the past. Now the species-diverse woodlands that make up South Texas are attempting a comeback, and viewing that great return is but an exercise in watching individual plant types return. Herbaceous plants and grasses mark the beginning of the great homecoming as do smaller hardwoods. Thus I'll abruptly encounter guayacan, leather stem, amargosa, coma, and desert yaupon, and in my mind's eye I'll see the sun's energy flowing through the plants cleanly and efficiently. The transition zone marks that area, as the name suggests, where one ecological region merges with another. In some places these regions are influenced by temperatures, rainfall amounts, and available sunlight. But the Sand Sheet is an entirely different phenomenon. No less destructive than a tsunami that sweeps inland to drown the earth, the Sand Sheet (though a much more gradual process) blew inland and concealed all that had gone before.

But even as nature attempts to fix things there are those who are either unwilling or unable to understand the process. Destroy the plants and you ultimately destroy yourself. While the delta to the south was once a fertile admixture of organic materials blanketing the ground on either side of the Rio Grande, the Sand Sheet is but an aggregation of countless grains of silica unable to capture nutrients or water. Alkalinic with summer ground temperatures reaching as high as 48° Celsius, it seems forcing the sand to become farmland would not occur to people unless they had no other choice. And yet some have tried. People have told me that watermelons grow well on the sands. They generally do not include that one or two crops leave the area so depleted that nothing grows afterward and it takes years for the sand to begin its succession back to herbaceous shrubs and grasses. One evening I sat on a man's porch looking out at what had become the quintessence of wind-driven sand. He'd wanted to play the part of farmer and had pumped money and more money into the project with nary a sign of profit. He said, "See that brush out there," then pointed southeast. "Yes," I said. "Well," he continued. "I'm going to clear it so I can expand my farm out that way." I sat speechless for a moment amazed at what the man had just said. Finally, I noted: "If you clear that brush over there you'll have nothing to break the wind and you'll be covered in sand every time it blows." He didn't reply but the brush was, thankfully, never cleared.

But the native plants have learned to coexist with the habitats in South Texas be they classic brushland or in this case the sands. My impression of the Sand Sheet in the beginning was that it's basically one dimensional. Gone is the incredible diversity one sees on those places that have not been disturbed but instead allowed to mature as nature intended. Of course, since the tsunami of sand occurred only eleven thousand years ago and has crept inland since then, one sees a reduction in diversity. Indeed, the motts that represent a foothold on developing ecological synergy are themselves primitive if one examines them from the point of view of variance or divergence. As mentioned earlier the western side of the great Sheet has motts consisting mainly of granjeno, brasil, and mesquite. The eastern side has motts of live oak.

But the transition zone, like a kaleidoscope of plants, always makes me smile. And leather stem is an icon of classic brushland. It has other names: dragon's blood, *sangre de drago*, *Jatropha dioica*. When you slice into the thin cuticle the fluid underneath bleeds a crimson red. As a boy I learned the blood of leather stem was used to treat mouth ailments and some claimed it cured kidney infections. A *curandero* (healer) named Julian said he used leather stem for sore throats as well. He said he'd cured diabetes and pneumonia and scores of other illnesses with nothing more than the plants growing wild in the brush.

Keeping track of plant species' abundance and distribution helps one establish the border between the Sand Sheet proper and the transition zone. As organic matter is added to the ground more species gain a foothold. For that reason, the Sand Sheet is a land in flux. What you find today did not occur yesterday and what you find tomorrow will be different from today. For it can be said that all plants on the Sand Sheet proper are recent arrivals. Everything you find had to rearrive and as such, no one can correctly say this or that is native but that over there is nonnative, for those terms are nothing more than perceptions oblivious of evolution's unyielding march. Fifteen thousand years ago and there was no Sand Sheet. Regardless, parts of the Sand Sheet now appear stable while other areas continue creeping outward. It is entirely plausible that given ever-warming conditions and persistent droughts the sand will advance farther northwest as the plants that hold it at bay succumb to the heat and the lack of water.

I sat watching my campfire and waiting for rain and then decided to make a wooden spoon with my hook knife. Like most of my crooked knives, I make hook knives using small mill files. It's perhaps not the best method since I'm not sure what type of steel I'm using, though I suspect it's one of the tool steels like W1 or W2. Regardless, I've never encountered a problem using old mill files, so until I run out of my stash I'll continue making crooked and hook knives with files. Instead of incorporating several angles in the blade and handle as in the crooked knife, the hook knife, as the name implies, is shaped like a hook. The hook knife uses the same chisel bevel seen in the crooked knife. I bevel my knives on the inside of the hook and then while still in the annealed state gently bend them into whatever style of hook I need.

To make a good spoon requires a slightly curved branch so I sawed a suitable piece of mesquite. Unlike northern hardwoods that tend toward the soft side, Texas hardwoods oftentimes feel like rock and will dull a knife quickly. When fully dried, woods like mesquite, ebony, Wright's acacia, chaparro prieto, and huisache are not only brutal on knife blades but also can strain the ligaments and tendons of the hand and wrist. Super hard South Texas woods like guayacan and brasil are so dense they won't even float in water.

After I've completed the initial shaping of both the spoon's bowl and handle, I begin using progressively finer grits of sandpaper starting at 80 and finishing at 600. On that day, however, I used my pocket knife's spay blade as a scraper to smooth out the bowl and the handle. Before beginning the project it's a good idea to coat the ends of the branch with a sealant to keep the wood from drying too quickly and cracking. In this case I used the sap from nopal. I'd also collected a forked branch onto which I attached a small adze blade I'd forged years before. I prefer an adze handle with a fork of about 50 degrees, the handle piece twelve inches long, and the section where the adze blade rests four inches long. Using my pocket knife, I whittled out a shelf and placed a leather pad over it then fastened the adze blade with tarred nylon seine twine. Using the adze, I chopped the mesquite branch into the rough shape of a spoon and then began carving the bowl with the hook knife. Green or wet wood makes the work easier.

It's not hard to make a hook knife but one should know a little about the properties of steel. After heat treatment I temper the blade to about 59–60 Rockwell, thus making for a well-hardened knife. At the juncture of blade to handle I temper down to about 50 Rockwell in order to alleviate stress at that point.

I place the spoon-to-be in the palm of my hand, then using the hook knife gently shave crosswise from one side of the bowl to the other until a shallow pit is formed. I don't go too deep and I watch for tiny checks at the tip of the spoon and the connection of the bowl and handle. If the bowl is carved too deeply it becomes uncomfortable to use, but a scoop or ladle requires a deeper bowl. It took about thirty minutes to make the spoon and thus another circle was completed. Ate my sandwich, took my morning medicines, drank my coffee, and waited for the rain that never arrived.

I placed the spoon I'd made into my musette bag and decided to set off on a contemplative walk into nature. Found a thick covering of trees within a long mott and then sat on moist ground, the stillness emanating around and within. "There comes a time," I whispered. Others had come to this very spot before me and someday someone else might arrive and discover what I and those before me found. A tree delivers its branch for my spoon and at other times for my bow and walking stick. But then the tree, as if to say, "I will," sprouts four or five or ten small branches from where the branch was originally cut and each sprout grows straight into the sun. It would not be so had I chopped down the tree, but a coppicing allows the tree to send a sortie of branches upward. Life renewed.

I recalled two badger holes not far from where I sat. I'd first spotted the holes a couple of years before when I moved to the Sand Sheet, noting that the badger had abandoned the holes several months previous. The two holes were also in the transition zone between Sand Sheet and Brushlands, and at the time I saw no loose dirt or fresh tracks by the holes and neither did I find recent droppings suggesting another animal had used either hole for its burrow. During the ensuing months I noted the holes were beginning to fill. So that morning I decided to walk the quarter mile or thereabouts to the holes to see how they now looked. The north wind persisted but I was walking south by southeast and so the winds were at my back. Soon I entered an area dotted by stunted mesquite trees amid a ground etched by shallow gullies and scatterings of limestone pebbles, the detritus from larger boulders squeezed upward from the underworld. The mesquites muffled the wind and the silence returned. I saw the place where the holes had been dug near a snaking groove cut deep into the ground by cows. But the badger holes were gone or at least nearly gone. One filled completely and the other a mere depression in the earth. No one but he who saw it in the beginning would know what had been there but a few years past. Who will write this history? I wondered. Who will tell the story of that which came without power and wealth and that won no wars or built great cities or founded a popular religion or authored a seminal book or solved a great riddle or invented a remarkable machine? And yet, those two badger holes, one now vanished and one nearly filled, told a story about life in perhaps as intimate and meaningful a way as have all the other stories we have come to know.

I stood a moment then walked back to the main trail because to continue through this scantly covered spot meant I would reach a place where everything that was is no more. Once upon a time deeply wooded ramaderos traversed the landscape between subtle rises in the ground. Old growth mesquites and ebonies grew in riparian belts along the ramaderos, and forests of chile del monte, granjeno, coma, and chapote intermingled with vines and mossy mats that stretched along the ground and into the trees like green and gray carpet damp to the touch. Now it was a place where introduced grasses grow only when it rains. For most of the year the land lies barren. So I turned and faced the north wind and walked into a thin patch of woods. It began misting and the trees grew solemn and quiet. Spider webs nearby were dotted by watery crystals. Sunlight cut a thin line through the clouds in the west and for an instant turned each web into a rainbow. All this is but a remnant of what lived here. The woods that once flourished, like the badger holes dug years ago, cannot be imagined by those who wander by now and see only flat earth.

12

I hunted as a boy and young man but then one day I stopped. Does a glass fill or run dry? I have no idea, but hunting no longer appealed to me and it was as if someone else had taken the place of the man that used to be. I know men older than I am who remain passionate about hunting. They are not lesser or better men but instead different men. But it seems to me that as a hunter matures he looks at hunting differently. The day of the high-powered rifle is replaced by the bow and if the man were younger and stronger he might opt for a spear. Regardless, I never hunted on the Sand Sheet and in fact had not hunted for over a decade. I'd stand on the back porch shooting my selfbows and carrizo arrows with projectile points made from cow bones or band saw blades, but with the exception of a wild hog that went after one of my dogs I pursued nothing more than bales of hay. I did, however, think about how early humans hunted. In the beginning they stalked beasts within groups linked to technologies that required cohesiveness. If ceremony existed beforehand or afterward it came as pleadings and then gratitude to whatever made the acquisition of food possible. Even so, hunters often risked their lives because successful strikes meant approaching the animal closely. Ultimately, however, even atlatls proved inefficient after the great megafauna disappeared. Smaller animals like deer could easily spot the arm and hand movements needed to propel an atlatl dart, and so a new technology was needed. But the bow's arrival ushered in much more than a better way to obtain meat. It also freed the hunter from the group. How this occurred remains open to hypothesis, but most likely the initial event leading to the ability to hunt alone came when someone realized that a cord made from animal or plant fibers tied to both ends of a branch or sapling could propel a long stick from one point to another without much effort. We can assume this discovery was a serendipitous event and might have happened many times before someone connected the phenomena to hunting. As often occurs in life it was not the event that proved epiphanic but instead its interpretation. But the technology already existed for making exquisite spear and atlatl points. In addition, humans had previously learned the value of flight stabilization and its relationship to trajectory. Affixing stone and bone points to arrows and attaching fletching was almost certainly added early on and long before the rudiments of bow tillering and pull-weight determination were achieved.[1] Nonetheless, the few ancient bows that have been discovered were tillered as well as any made today, and the technology for making exceptional bows has improved only slightly in the following millennia. Modern mass production has given us fiberglass bows and, the ever-present need to make things easier has led to the invention of apparatuses with pulleys, cams, and cables called "compound bows" that are only distantly related to the classic and true bow. To that end, the ancient selfbow remains the quintessential endeavor of the toxophilite, and for those enamored with traditional archery no other bow exists.[2]

I had long since wondered what hardwoods were used to make bows by Indians living near the Sand Sheet and within the delta region to the south. That quest continued during those years at our secluded home. In South Texas trees seldom grow in straight lines and I searched for suitable bow staves like others might hunt for precious gems. Such staves grow in hidden places blanketed by shade and concealed in nearly impenetrable thickets where competition for light forces plants skyward. But which South Texas woods made the best bows remained a mystery until I began experimenting with the regional flora. My exper-

iments began in the early 1990s and continued once my medical issues improved. There are no real experts on South Texas Indians, for what little information we have comes not from the Indians themselves but through the eyes of Spanish explorers. The Indians were an afterthought, an obstacle to circumvent, a curiosity to most and a potential pool for conversion for others. But even the Roman Catholic Church that was firmly in control of the Spanish monarchy sought wealth and power more than converts, and, in fact, did not even consider Indians in possession of a soul and thus worthy of redemption until nearly one hundred years after Columbus's arrival in the New World. Even then the indigenous people were treated more as objects useful to work the mines than as equals in the presence of God. Regardless, good Spanish archival data on the Indians of Deep South Texas and northeastern Mexico comes from an anthropologist named Martín Salinas. But even Salinas's work contains only a scattering of bow and arrow documentation.[3]
Bows were used for hunting, fishing, and warfare, but the only noteworthy description of regional bows comes from Ladrón de Guevara who in the mid-1700s mentioned the bows used by Indians in the San Fernando River region of Tamaulipas, Mexico, a land I am well familiar with having spent a good part of my youth in the 1960s exploring and hunting on my father's ranch, El Cuervo, that bordered the San Fernando River. Like the South Texas Brushlands the area around the San Fernando River was formerly a world of thorn trees like mesquite, ebony, chaparro prieto, and granjeno, along with numerous cacti. It also included a member of the citrus family called *barretta* that is rare in the United States and found only in isolated parts of Starr County, Texas. Guevara said the Indian bows he examined were the length of the user, or about five feet.[4] The Spaniard also noted that Indians carried arrows in quivers, but Salinas

makes no mention of whether or not Guevara says the quivers were made of leather or in the form of woven baskets. Furthermore, Guevara said the Indian's arrows were constructed in two parts, which was a familiar method of making arrows in various regions of North America. The shaft, according to Guevara, was made of *carrizo*, or reed. He said the arrow points were made of either flint or glass and the foreshaft was made out of "heat-treated hardwood." No details are provided regarding what type of wood the Indians used for arrow foreshafts. In the case of the Indians along the San Fernando River, the glass used for arrow points was scrounged from the Texas Gulf Coast about fifty miles to the east in the form of bottles or shards washed ashore. I wonder what the native people thought of the glass they found. Guevara also noted the Indians used asphaltum that likewise was swept onto the sands. This black rocklike material comes from oil oozing out of fissures on the Gulf's floor. When heated it melts into a tar-like adhesive ideal for bonding stone and glass projectile points to the shaft. The Spaniard does not say whether or not sinew or some other fiber was used in addition to the asphaltum glue. My own experience making hundreds of carrizo arrows suggests that cordage made from agave (*Agave sp.*) was most likely preferred over sinew, especially for arrows and spears used for fishing since sinew softens when wet. The high humidity along the coastal areas would also preclude the use of sinew for that same reason. It's improbable that hide glue was used on archery equipment as it too is water soluble. In other areas of North America pine pitch or glue made from fish skins was used to attach arrow points. But in South Texas and northeastern Mexico the Indians found a near perfect bonding agent in the form of asphaltum. Salinas does not say whether Guevara mentions how many feathers were used to fletch the arrows. Was it three or two? We can assume the Indians also applied asphaltum to affix feathers to carrizo shafts. In addition to asphaltum they may have also secured feathers with thin strands of agave fiber or some other plant cordage. Another adhesive used in South Texas is mesquite sap glue, but since it is water soluble it's unlikely it was a preferred binding material on arrows. My grandmother made mesquite sap glue for my mom and her siblings' school projects when they lived in El Centro in the northeastern corner of Starr County. A clump of mesquite sap known locally as *chauite* (*chow-wee-teh*) was dropped into a jar of water where it quickly dissolved. The children dipped their fingers into the jar and then dabbed the glue onto whatever paper they were using. But for arrow points, especially those that might be used for fishing, mesquite sap falls short. The people of the Coahuiltecan Geographical Region knew from experience what worked best in their area.[5]

I have no doubt that Guevara's use of the word "reed" or *carrizo*, refers specifically to a plant found in abundance along waterways, wetlands, and inlets in South Texas and northeastern Mexico. *Phragmites australis berlandieri* is a subspecies of a plant encountered worldwide known as *Phragmites australis*. The other common subspecies in North America is *Phragmites australis americanus*. In pre-Columbian times (before 1492) *Phragmites* presumably occurred in North America as those two subspecies. Over time, however, other subspecies of *Phragmites australis* arrived as a result of what has become known as "The Columbian Exchange." In effect, European exploration and subsequent settlement initiated a transfer of plants and animals in both a western and eastern direction. Potatoes, tobacco, cotton, and corn went to Europe, while wheat, barley, apples, and almonds came to America. The exchange included hundreds of species. Europeans gave America rats, pigs, horses, and donkeys, whereas America gave Europe turkeys and llamas. America received chicken pox, measles, leprosy, and smallpox, and Europe was paid back with syphilis and Chagas' disease. The list is long and complicated and as other parts of the world journeyed to America the list grew expansive. But the Columbian Exchange was more often than not unintentional and thus the European subspecies of *Phragmites* eventually took hold on American soil and in the intervening years has become "invasive" in many wetland areas. The word *invasive* is a popular term within both scientific and economic circles when referring to *Phragmites* and other opportunistic plants and animals. Efforts are ongoing to destroy *Phragmites* in many parts of the country. Perhaps, however, it would be wise to consider plants like the European subspecies of *Phragmites* less of an invader than a messenger. In opting for that point of view we might then find better avenues for dealing not with the plants themselves but instead with the specific human behaviors that predispose the expansion of these opportunistic species. As human populations skyrocket, for example, so do the levels of soil disruption and aquatic sedimentation that favor one plant over another. Employing massive quantities of artificial fertilizers also creates an imbalance. Increased salt inflows from roads and industrial runoff likewise provide opportunities for one species while reducing the survival of another. Thwarting the expansion of plants like *Phragmites* has to date revolved around the use of potent herbicides that often cause even more problems. Herbicides can devastate various microorganisms that are vital to a healthy soil matrix. Herbicides are likewise harmful to sensitive individuals who might breathe the fumes or in some way come in contact with these poisons. These chemicals also eliminate pollinators that are already diminishing in most places, and they kill many amphibians in wetland areas. Note however that the various *Phragmites* subspecies look essentially alike and differentiating between them requires an understanding of both morphological and genetic variances. But journey back to America's Neolithic past and we find that *carrizo* was viewed not as a menace but instead a blessing. In the Coahuiltecan Geographical Region, indeed all along the edges of the Sand Sheet, these hardy reeds were used to construct dwellings called *jacales*. The roofs were thatched with *carrizo* and the long slender reeds were crushed then blended with mud to form the hut's thick walls. The Indians made rattles, flutes, drinking straws, and pipes from *Phragmites*. Parts of the plant were eaten and a sugary exudate formed by aphids living on the reeds was also consumed. We now know that the Spaniard Ladrón de Guevara saw but the tip of the iceberg when he observed the Indians near the San Fernando River using *carrizo* for their arrow shafts, for the reed was as much a mainstay for the Indians of the region as was the *nopal* and the mesquite bean.

In pre-Columbian times however, the Sand Sheet offered poor pickings for the bow hunter. Deer were available but nowhere in the numbers found along riparian belts to the south and north. In recent times the wild hog has made an appearance, but they have taken hold only around manmade ponds or other artificial water sources. I asked a landowner named Steven Burns if he had any problems with hogs. "Very little," he said. We were standing atop a magnificent migrating sand dune and looking into the horizon. The sand-ridden plain stretched in every direction with only feeble scatterings of hardwoods here and there and dried patches of shrubs and grass in between. Far to the southwest I glimpsed a lone windmill, but otherwise the seemingly endless expanse appeared void of life. "When there's no water there're no hogs," Steven said. Sometimes I'd spot the black figures, some of them quite large, grunting and snorting but always keeping close to the motts and seldom trekking across open areas except at night.

I've often wondered if the bow and arrow provided not only a means to roam the woods alone but also created an opportunity for contemplation. In a group people remain wedded not only to the dictates of the many but to their thoughts as well. But in isolation our thoughts are allowed to run free. It is the time to dream and analyze and perceive the world in ways not easily accomplished when in a crowd large or small. As I roamed the motts I often imagined those who might have journeyed this far north from their enclaves along the Rio Grande. Were their bows made of cedar elm? That seems the most likely material to have been used. Perhaps they carried bows made from anacua or chapote? Despite other stories, it seems unlikely that mesquite or huisache or chaparro prieto or ebony was employed to make bows if the woods mentioned above were at hand. Those who used inferior woods did so not by choice but because they were forced to use them. Perhaps, however, there was a trade route that somehow circumvented the sands or at least allowed travelers to traverse the region without fear of dying of thirst and provided them as well with materials to make bows, arrows, and projectile points by which to hunt if perchance the items they brought with them were lost. Maybe there was a way to acquire pecan or ashe juniper or even Osage orange from the north? As the months went by a theory began to develop, though I was not yet ready to accept it completely.

Not a day went by that I didn't have work of one sort or another around the house. I tended the garden and kept the grass mowed to spot rattlesnakes in the yard. At night I often spent hours in my little shed making knives or bows and arrows. Sometimes I'd make spoons or bowls with my crooked and hook knives. Surrounded by a nighttime darkness few people know in these modern times living as they do beneath ever-present street lights and neon signs, I'd peer into the dimensionless black beyond the shed listening to the faint whistles of elf owls or the melodies of an insomniac mockingbird. If my dogs began barking, I'd grab a flashlight and walk out from underneath the tin roof and shine the light down my little driveway or into a line of trees marking two giant motts to the west. Inside the cabin I'd sometimes play my guitar or read a book. But too often an hour or two before sunrise I'd awake abruptly and think back. The months of debilitation and hospital stays seemed far away, but every three months I'd visit Dr. Speeg in San Antonio and each trip was preceded by a level of anxiety. So I'd spend nights working in my shed or walking the hushed woods as a means of distraction. Regardless, thinking back on all the days gone by becomes, whether we like it or not, an integral part of growing old. My ruminations often took me back to a man I once knew who told me, "When I see a piece of land the first thing I think is how I can exploit it." That man meant a lot to me and I still remember my feelings when he made that remark. I realized we were as different as any two men can be because my first thought when I see nature is always how I can preserve it. I believe we are born not to take or plunder or exploit but instead to learn and hold in reverence those things that cannot speak for themselves or fight for their freedom.

Some years back a graduate student from a university in New England interviewed me. She brought up the subject of immigration, and when speaking of the people who cross the Rio Grande into the United States she said, "They're the victims." I shook my head and replied, "No they're not." Then I added, "Who speaks for those that have no voice? Who speaks for the palo blanco tree or the bald cypress destroyed along the banks of the Rio Grande? Who speaks for the tortoise choked to death on plastic discarded by those same people who swim to the US side or the bobcat or jaguarundi poisoned by toxins leaching from tons of trash abandoned at the river's edge? Those are the true victims! Who speaks for them?"

Like many times before I found a quiet spot and looked westward into a bank of clouds that reached north and south obscuring the setting sun. I watched as the colors changed within the clouds. In the beginning I saw yellows and then orange. As the sun dipped lower toward the horizon long rays pierced the sky over me and then disappeared to the east. The colors darkened into deep pink and then red. I thought back over the years and all the sunsets I've seen sitting in quiet places too numerous to recount or even remember. In a few minutes the sky went from red to gray. A screech owl began trilling from somewhere to my right. A few minutes later as the tattered edges of the gloaming faded into blackness a gray fox walked out not more than thirty feet away. It never saw me as I sat motionless watching it sniff and scratch at the ground. And then it was gone. "For you," I whispered.

13

On the afternoon my son Matthew drove up to the house toting a U-Haul trailer behind his pickup truck I was sitting on a rocking chair on the front porch. I'd spent every late afternoon for the past two weeks in that rocker hoping that on that day Matthew would arrive. The sun would begin setting as darkness crept over the mesquite trees in the yard, and another day would pass. I'd walk into the cabin thinking that maybe tomorrow would be different. The next day I'd tend to chores while glancing down the little dirt road leading to the gate. Near sunset I'd once again sit on the rocker, eyes fixed on the driveway. But another night would come and I'd go inside, eat supper, and then go to bed. One day became the next but there I sat watching and listening and waiting. Then as if every good thing in the universe had been bestowed on me on a sunny afternoon, I saw Matthew's truck in the driveway, and as he stepped out and walked toward me he said, "Hello, Dad." I'm not sure if I said anything but I know I reached out and hugged my son and what I wanted to say was, "I love you. It's so good to see you." At that moment it was as if the world around me had disappeared and all I could see was Matthew's smile and all I heard was his voice. Matthew was going to live at the cabin, and it seemed too good to be true. As it turned out I needed to leave the very next day to see Dr. Speeg in San Antonio. So on my way I dropped off the U-Haul in Premont, Texas, about sixty miles north. Perhaps it was best to let Matthew get used to his new house and the woods surrounding it on his own. He was twenty-five years old at the time and I'd taught him and his brothers the ways of nature since childhood. Upon arrival in San Antonio I stopped at Central Market on Broadway for a sandwich and cup of soup.

I joke that you know you're getting old when your favorite place in a big city is a supermarket. But HEB Central Market had become my special place. Quaint without that sterile look of a modern building. There were always people shopping and eating in the two dining areas. College students and professors from Incarnate Word University next door and men and women in US Army uniforms from Fort Sam Houston nearby as well as patrons from the surrounding area. After my sandwich and soup, I'd have a cup of coffee—just regular coffee, American coffee, nothing fancy or sweet or with some exotic name but just good old coffee—and I'd sit looking out into the parking lot feeling at home and watching people come and go. I'd been to San Antonio so many times and always eating at Central Market that I knew the faces of practically all the workers. But I never said anything. Never visited with anyone. I'd sit in the eating section at a table where the light was dim enjoying my soup and sandwich and keeping to myself. Fortunately, Norma had come with me so we ate and talked and enjoyed each other's company. There was always an apprehension that things were going to change and, of course, they will someday. They always do. But on that visit like the others preceding it my tests were normal. Rarely are changes for the better though in my case I'd gone from very ill to nothing more than simply getting older. Trifocals, bad knees, the aches and pains of age but aside from that I was okay. Taking my medicines and trying my best to eat right. Walking every day if it wasn't too hot or muggy. Grateful for nature and the wilds surrounding me. Whether brown or green, it didn't matter. And yet I've known people who are as blind to nature as one can get. They spend their lives in a kind of stupor. Never stopping to contemplate the beauty around them, they exist angry and self-centered, and there is nothing anyone can do to help those kinds of people. What I've learned on this adventure through life is that one must pick up and go forward. Not in self-indulgence or vanity but instead in tranquility and contemplation and above all with respect for life. Even so, we discover purpose and understand beauty when we do not force it upon ourselves. We obtain healing by allowing the powers beyond us to fuse with the power within us. We achieve harmony when we distance ourselves from the bigotry of humankind and follow the will of nature. We perceive truth when we accept and revere the one thing known to be real, and that is life itself.

When we returned to our little house I was almost expecting to discover it had all been a dream. But Matthew walked out to greet us, and now I think of Matthew's homecoming as my second beginning along the Sand Sheet. I had my son with me and not only did I feel more secure but I also had the very special person who when he was little shared a cabin with his mother and me at a ninety-acre place called El Ponderosita. I'd take his hand and we'd walk the secret trails around the Good Earth Cabin, and that's what we called our home. But like so many other things in my life that place is but a memory. People arrived soon after we left and bulldozed the woodlands then created a world of houses and pavement with noise and lights and called it progress.

"How were things while I was gone?" I asked.

"I hung a hammock from those trees over there," Matthew said. "While I was reading I heard something below me. I took a picture of the rattlesnake. It was impressive."

With Matthew came Little Boo, a Maltese his older brother, Jason, had given him a few years before. So now we had four dogs, Pita, Oy, Maggie, and Little Boo. She was an inside dog but as good a sentinel as were my blue heelers outside.

A few months after Matthew came to live at the cabin we

were informed that people planned to drill for oil not far away. Some of the older folks from San Isidro said all this was folly and amounted to nothing more than another insult upon both the land and those who'd endured the decades of ruthless drilling and desecration. "There's no more oil out there," Tololo Lopez said one afternoon. He said his brother Omar had worked the fields when he was younger and had told him the land was all drilled out. But those who arrived to search for oil did not see the land as any special place. The drillers stripped the ground of every tree and shrub and flattened it with machines until it was like a skillet onto which they lathered caliche like lard in preparation of their feast. Then they brought in more machines and finally a monstrosity of a drilling rig that was erected and lighted and turned into a bullhorn that blared out anguish for a mile in all directions and from which the earth was pierced and poisons injected—xylene, toluene, benzene—and the night's sky was aflame with what seemed an evil glow and the air smelled of things rancid and through all hours of the day and night trucks drove passed our home without an ounce of thoughtfulness for the people who lived at the end of the winding driveway. But when a rapacious spirit comes your way then you must excise it promptly, otherwise it will bleed out all its surroundings and not once ponder the consequences of its own greed. And so on a cloudy afternoon we blocked the little road, and when the drillers drove back our way two fellows wearing hardhats and a startled arrogance stepped out of their truck and one of them said, "What the hell is this?" To which I replied, "How would you like it if people raced in front of your house with no concern whether it was blowing dust into your yard and not caring how it might harm you?" The man looked at me and I could tell his pride was getting the better of him, but I persisted. "You will respect us," I said. But the well produced nothing as had been predicted and the men in their hardhats and dingy coveralls abandoned the site, leaving a five-acre scar and a great black pool of oily gunk that hardened and then crusted over as if it was blood drying in the sun. A couple of months later they covered up the pit as if that would make things better. They bulldozed the caliche into one corner and then disappeared. The trees that had covered the land were gone and no one but the most naive would think the place was left in a natural state. But after my warning on that cloudy afternoon they drove past our house slowly. We were not to be mistreated and we would not play their game. Who are these people I wondered? I'd see them around San Isidro at noontime if I happened to be in town. I'd seen them speeding back and forth on Highway 285 between Fort Stockton, Texas, and Artesia, New Mexico, where the air is fraught with poison. When I questioned a convenience store owner in Artesia about the smell, he replied, "That's the smell of money." But as I walked back to my truck a young mother who had overheard the conversation approached me and said, "Sir, that's not the smell of money; it's the smell of death." I had seen them as well when I drove up FM 1017 from La Gloria to Freer at the edges of the Eagle Ford Shale Region where men without conscience lay waste to everything for miles around. Without shame or guilt. And I recalled those words that bit into me when I first heard them, "When I see a piece of land the first thing I think is how I can exploit it." So that's who they are and no waving sign or marching parade or singing of songs (*This land is my land, this land is your land*) or long-winded speeches or face paint and costumes will deter them. They make the rules bought in cash and enforced by those who do as told if it gives them power. Respect, I said. *You will respect us.* I promise you they will understand when they see the resolve on your face and hear it in your voice and realize you will act if pushed far enough.

14

The Sand Sheet occupies the northeastern edges of the Coahuiltecan Geographical Region. But while the Sand Sheet is real, Coahuilteca is imaginary, mapped out by European Americans to include a stretch of what is now South Texas and northeastern Mexico from about Del Rio to near San Antonio then east by southeast along the northern edges of the Sand Sheet to nearly the Gulf Coast. The region plunges south across the Rio Grande delta into Mexico and finds its southern border near the present-day city of Victoria in the state of Tamaulipas. Coahuilteca then spans westward to what is now Monterrey in the Mexican state of Nuevo León. No Coahuiltecan tribe existed, nor any specific common language or perhaps even dominant culture. The name was derived from the Mexican state of Coahuila that borders the United States south of the Texas cities of Eagle Pass and Del Rio.[1] The region was made up of hundreds of small family bands, each independent of the other, that roamed along waterways and stayed near wooded land where fuel to burn fires and materials to build shelters, and animals to hunt was available. The people persisted as hunters and gatherers until they were at last assimilated into the greater population of both Spanish and adjacent Indian groups. Some maintain that those who lived in the Coahuiltecan Geographical Region were backward and even slovenly but I reject those assertions outright. The region is one of the most inhospitable in North America, known by many as the Thorn Forest, where except for a few rivers and a scattering of wetlands it is eternally dry. In Coahuilteca families staked out their land to forage and hunt and then stayed there from generation to generation confined by the presence and pressures of neighboring groups. Only the hardiest and most resourceful people could survive under those conditions. Competition for water and resource-filled land was often intense. Most people dwelled along a river that today is called either *Grande* or *Bravo* depending on whether one lives north or south of its course. But to suggest the people subsisted in a backward sense is perhaps nothing more than the insertion of bias nurtured as much by personal values as by an unwillingness to accept differing cultural ideologies. Some scoff at the notion of the people eating insects and small reptiles, for example, and infer this meant desperation to find enough food. But as equally likely were societies not limited by the same taboos held by many contemporary observers. Regardless, throughout the history of human settlement along *El Río* people crossed in their canoes and rafts to join family members living on either side. The river was never a border but instead a communal waterway and trade route from which to fish and drink and likewise travel toward mountains in the west and the great sea to the east. But the desert in the north was always the mighty barrier, a moat made of sand, a flat and waterless region unfit for any sort of permanent human habitation. Even so, by the time European settlers arrived herds of wild horses and cows ran free across parts of the Sand Sheet, and from that Europeans extrapolated it was a place to build haciendas and cattle empires. *Porciones*, or land grants, issued by the Spanish monarchy established extensive landholdings that included much of the eolian sands. One in particular bestowed to a man named Cabazos exceeded a half million acres. But by now the great Indian horse cultures had likewise arrived and the earliest Spanish colonizers faced a nearly continuous onslaught of Apache and then Comanche raids. One story told in my mother's family was of a young boy, perhaps my grandmother's great-uncle, who was kidnapped by Comanche warriors. Years later he managed to escape and was able to find his way back to his relatives and his own mother and father. When he approached their house his siblings, now in their teenage years, surrounded him warning him to back away. He had long curly reddish hair but wore the dress of a Comanche. "What do you want?" one of his siblings asked him. "Don't you know who I am?" he said. "Go away," they told him. At that moment his mother walked out of the house and when she saw him she ran toward him and hugged him and said loudly, "My son, my sweet son." Some weeks later several Comanche warriors approached the house, and the young man's father and his brothers told him to hide. They held guns, percussion locks I assume, on the Indians but the Comanche were not afraid. "We are looking for the one who ran away," they said. "There is no one here like that," the young man's father told them. But one of the Comanche sniffed the air and replied, "He is here." At last the family persuaded the warriors to leave, warning them that if they returned there would be bloodshed. For many months the family did not let the young man out of their sight. But the Comanche never came back. Stories similar to the one told by my grandmother abound. In one of my *Woods Roamer* blog posts I wrote that nearly every old-time family in the region has a story or two about some ancestor who was kidnapped by either Apache or Comanche. In some cases, people who'd been abducted were able to escape years later and make it back to their families. Many others never returned. But even when children managed to break free they are said to have never fully acclimated back into the European lifestyle. Some came close but none of them ever forgot their Indian ways. It seems "the Christian life" lacked the adventures of the Indian way.

Raising cattle on the Sand Sheet proved challenging because grass growing on sand is tenuous and fragile. I learned that when I first built our house. The man who grazed cows on the land for two years before we arrived disrupted the area's delicate balance, and it took over two more years before it began to heal. Even then some people seemed unwilling to walk across any piece of land no matter how small, and the ruts from their trucks remained as scars for years. Driving along FM 755 between La Gloria and Rachal I saw vast stretches of land severely overgrazed. Oil and gas wells, recharge stations, crisscrossing roads, pipelines and cattle; for many, the Sand Sheet means nothing more than something to exploit. But others looked at the land in a different way. To them it meant home; it meant sanctuary. When a pipeline company crossed a family's land west of us, they were outraged. Such are the politics in Texas where the oil and gas companies control all levels of the government from so-called regulatory agencies to seats in the legislature and perhaps even higher. And so as the rules go, gas and oil companies can stake out pipeline routes as they please, landowners and citizens be damned. I talked to one of the family members whose land was cut open for a pipeline, and the word *resentment* seems tame when remembering that conversation. Another man I spoke with lived in that metropolitan chaos to the south called the Rio Grande Valley. He said, "The Valley is running out of water and we figure we can dig deep water wells here on the Sand Sheet and pipe the water to places like McAllen, Mission, Pharr, and Edinburg." *And create even more chaos*, I thought. But then I reminded him that as we spoke billions of gallons of underground water were being fracked (hydraulic fracturing) in gas and oil drilling. The fracking fluids contain from five hundred to eight hundred carcinogenic and toxic chemicals like methanol, ethyl-benzene, naphthalene, and hydrogen fluoride.[2] "Most of those chemicals are left behind to pollute the underground water," I said. "In fact, they recover less than 40 percent of those fluids which are disposed into other wells—the same kinds of wells causing earthquakes in North Texas and Oklahoma. So when you drill for water on the Sand Sheet what you'll find is something too polluted to use." The man I was speaking with had no answers and neither does anyone else other than to stop the madness and pray that too much damage has not already been done.[3] Later someone told me of a man who was proposing that water containing fracking toxins be used for drinking after the water had been "processed." Mindless desperation, I thought. We take pure water from aquifers and then saturate it with poison and then we decide we want it back but first we must extract the poisons we purposely added.

But on our homestead we tried our best to maintain a distance between those who saw the land as something only to exploit and nature as nothing more than an obstacle to be circumvented. I continued with my quest to learn the favored hardwoods for building bows by those who lived in the Coahuiltecan Geographical Region, and I became increasingly convinced that the earliest European explorers to the area had no desires to understand native bow technology and only a passing interest in things like native languages and cultures. Over time the picture became more and more clear. The regions bordering the Sand Sheet harbored great bow woods that the people no doubt used, though those woods are not mentioned in the historical literature. The eastern Sand Sheet holds one supremely efficient bow wood, live oak, and also cattails from which one can make suitable arrows. Nonetheless, the scarcity of water would have made travel across the sand a risky enterprise. Even seasonal rainfall would not have guaranteed they might find water along the way. Regardless, I reasoned that if a pre-Columbian north–south trade route was contemplated then it could only have occurred through the eastern edges of the Sand

Sheet hopscotching from ephemeral pond to ephemeral pond created by blowouts or the settling of layered sands over clays. Otherwise, traders would have been forced to follow a route far to the west in the area that became known as *Los Caminos Real* or the Comanche Traces.

Whether using bows and arrows or atlatls, the natives who ventured near the Sand Sheet brought their weapons on their backs. Atlatl darts are long and delicate and require appropriate wood, of which the best in the region is *Phragmites*. But *Phragmites* is not available on the Sand Sheet, and woods used for bow staves must be dried properly or they will weaken with use and become inefficient. While granjeno can be employed to make a reasonably efficient arrow on the western Sand Sheet it must be kept long and thin in order to reduce spine weight and successfully bypass the bow's grip section. We have no definitive proof that the bows made in Coahuilteca were of the D-bow design (bend through the handle), but it seems reasonable to assume that this classic profile was used and thus arrows were built to withstand the competing forces created at the bow's grip.[4] Although cattail arrows are fragile they will work in a pinch. Stone tools would have also been carried into the desert since the Sand Sheet is as vacant of suitable rock as it is of water. For those reasons it seems unlikely that natives spent much time hunting within the Sand Sheet and only minimal time hunting its borders. Remnants of transient habitation have been found in areas in close proximity to the Sand Sheet, but this does not suggest that encroachment into the desert was the norm. An even greater likelihood is that these ancient campsites that border the Sand Sheet occurred before the sands swept across the area to bury the lands beneath. Regardless, from a practical and logistical perspective it makes no sense to propose that the desert lured hunters onto its sands. The densely forested riparian belts along the Rio Grande to the south and paralleling the Nueces River to the north provided infinitely more practical hunting and foraging opportunities. Hunting amid the sands whether with atlatl or bow is far too energy consumptive to be worthwhile, especially for those on foot. One of the frequent comments made to me by long-distance travelers attempting to trek across the sands is that it wore them out. In fact, a young man named Rene said that nobody he knew had ever successfully walked across the Sand Sheet during the late spring or summer. He said he knew some who had tried but they always turned back to town. "So how do people cross?" I asked Rene. "Workers on the ranches have been paid to watch for them and to give them rides and water," he said. But crossing the great desert on foot, even in winter, will turn the most powerful leg muscles into mush after a few miles. Remember also that the Neolithic natives of the region viewed hunting, fishing, and foraging in a pragmatic sense, in many respects no different from modern human's trips to the supermarket. The object was to acquire food and not indulge in sport or fancy. As such it seems that hunting forays beyond the fringes of the most productive areas would have been the result of an extreme event and thus avoided at all costs. Compare this to modern hunters who throughout the fall travel into the Sand Sheet to hunt deer as well as exotic game transferred there for sport. Decked out in camouflage and driving trucks and ATVs they arrive with a singular mindset. We even saw huge private jets flying overhead en route to landholdings replete with great runways and hunting lodges. But take away their trucks and planes and their high-powered rifles. Take away their canned goods, ice chests, and electrical utility lines. Take away their binoculars and rifle scopes. Take away their deer towers and all-terrain vehicles. Make them build their own bows from a sapling or branch. Tell them they cannot hunt with anything other than what they have made with their own hands. Selfbows and river-reed arrows; projectile points made from rocks or bone; fletching derived from wild turkey and secured with asphaltum or cordage made from agave fibers; knives and scrapers knapped from chert; water containers fashioned from the stomachs of deer. And then force them to walk into the desert and wait for them to arrive. Wait and wait and then wait some more. Look out across the flats and see if you can spot any camps.

15

Rain in South Texas falls mainly in late summer and early fall and thus resembles a tropical pattern as opposed to the patterns found in temperate regions to the north. Sometimes hurricanes arrive but mostly nothing more than a series of anemic tropical waves sweep overland. But while flooding occasionally pesters residents in the delta region along the Rio Grande, the Sand Sheet seldom sees anything more than a transient pond that quickly fades, oftentimes lasting but a few hours and most certainly not more than a week. In some places ponds might remain for as much as a month or more, but those areas are invariably at the distant edges of the desert known as the "transition zones" where the sand is scarce and clay forms a nearly impermeable table underneath. In other places nearer the Gulf Coast ponds form as long as a nearly continuous supply of groundwater keeps the sand saturated, thus inhibiting drainage. Near our house we dug three small ponds. At the gray water outlet, I planted *Phragmites australis berlandieri* in order to grow my own arrow shafts. But carrizo also kept the gray water from smelling and the reeds inhibited mosquito growth. A few hundred feet farther west Matthew dug another small pond where we ran a continuous flow of water from the well for over a year. As long as water poured into the pond it remained full, but if we shut the faucet the water disappeared in about five minutes. This came to light one day when two professors from the University of Texas–Rio Grande Valley, geologist Juan Gonzalez and anthropologist Russell Skowronek, brought a group of students to drill boreholes to see how far one had to dig before encountering the red Goliad soil beneath the Sand Sheet. The two professors believed that small water-filled ponds could be encountered on the Sand Sheet, and that as a result ancient people had lived on the great South Texas desert. But I explained to them that was not possible. Perhaps there were small temporary camps along the transition zones but not on the Sand Sheet proper. "The sand sheet does not hold water," I told them. So as proof I turned off the water supply that had been dumping water into Matthew's pond for over a year. Then a couple of students set their stop watches to see how long it took for the water to drain away. In about five and a half minutes the water was gone. I remember the look on Juan's face as he realized the old man knew a thing or two about the Sand Sheet. "You see, it doesn't hold water," I said. Matthew had dug a third trough in front of the house, and that too remained full as long as water was pumped into it. I can think of no better example of the Sand Sheet's intolerance of ponds than those artificial and extremely ephemeral troughs we dug around our home. Sometimes a heavy rainfall delivered two or three inches of water in a couple of hours. One year our entire supply of rain for ten months arrived in less than two weeks. But as always the ground was dry the next day. I always told long-distance travelers there was no water for nearly fifty miles. "You will die of thirst. You won't make it." But for the most part travelers were infrequent. In other places I'd seen them walking in groups as large as forty in a bunch. But living on the Sand Sheet, we were for the most part left alone. Many of our neighbors weren't as fortunate. Some reports were alarming with tales of barns and sheds broken into and even houses burglarized and in some cases people assaulted and left for dead. We also heard about armed drug smugglers following trails through the brush, and reports in the media told of a growing instability in Mexico where criminal gangs and drug cartels battled each other for power. "What I've wondered is who is really in charge in Mexico," said my old friend Toni Treviño one afternoon when she and her husband, Benito, also an old and dear friend, visited the cabin. Toni was an assistant US attorney and her concern meant a lot to me. My father had owned brick plants in Ciudad Camargo, Tamaulipas, Mexico, and we'd heard those plants were now occupied by a drug cartel using the buildings to assemble armored vehicles. Friends of ours in Ciudad Camargo disappeared and the town itself was attacked several times by the cartels. All along *La Frontera*, bordering the United States from Matamoros near the mouth of the Rio Grande to Nuevo Laredo two hundred miles upriver, were reports of killings and battles between Mexican soldiers and militarized forces equipped and trained by the drug smuggling alliances. Ciudad Mier, a few miles west of Ciudad Camargo, became a ghost town when it was overrun by the warring drug factions. "You need to stay vigilant," said a friend of mine named Joseph Lopez who worked with a special unit of the US Border Patrol. He and others in the Border Patrol would sometimes drop by to visit. I looked forward to their arrivals because it was a chance to talk about the goings on around us as well as an opportunity to show my latest knives and bows to young men who appreciated those sorts of things. On one occasion several of them asked me about survival in the Brushlands and within the Sand Sheet.

"Sometimes I have to bail out of my vehicle pursuing people and it happens so fast I've seldom got time to grab anything," said one of the agents.

I replied, "There're five things you need on you at all times aside from your weapon and communications, whether radio or cell phone. You should carry a brightly colored bandana either yellow or orange in case you get hurt. Place the bandana on a bush so a helicopter can more easily spot you. Also carry a signal mirror and a sturdy pocket knife."

The agent asked me what type folder I used and I told him I was old fashioned in that regard. "I carry two jackknives," I said. "A carbon steel trapper model and a Swiss Army knife field model."

"So what's the fifth thing you need to have on you?" the agent asked.

"A small flashlight," I said. "Get one that'll fit in your pocket and make sure the batteries work or it's fully charged."

As they were about to leave I asked them how familiar they were with rattlesnakes.

"Can't stand 'em," another one of the agents said.

So I told them to wait and I went into my shed and grabbed four walking sticks I'd made from granjeno and then said, "Here's one for each of you. Take them with you into the brush and use them to poke the grass and shrubs checking for rattlers. They're also useful tools for tracking."

About two weeks later Joseph called and said the walking stick I'd given him saved him from an encounter with rattlesnakes on two occasions. Granjeno makes excellent walking sticks that aren't too heavy and yet offer the strength needed to test the ground for snakes. My favorite South Texas walking stick wood however is retama, which grows in low areas. It's lightweight though not as strong as granjeno.

One afternoon when Joseph was off duty he brought his father, Andres, to visit. Andres had worked for the railroad near Galveston but was now retired. Andres, Joseph, Matthew, and I sat under the tin roof of my little work shed talking about life along the Sand Sheet. About the tranquility and quiet nights and inspiring sunsets, and about the birds at our feeders and watering stations. We talked about the desert and the lack of water and about woodcraft and survival. We discussed those things that men who love the wilds share among each other when in places removed from the world most have come to know. Joseph was always concerned that we lived so remotely, though I think he knew this was the only kind of life for men like me and like him as well as he often mentioned that someday he hoped to live in a secluded locale. I sold Joseph two of my large chopping knives but I think I would have just given them to him had he asked, such was my appreciation of his friendship and willingness to check on us at our faraway spot in the great South Texas desert. But I too was concerned about Joseph's safety. Although most of the people who came for water were good and posed no harm, there were others who were clearly dangerous. One morning two men walked up to one of our faucets only a few feet from the front door and one began filling a water bottle. Perhaps they were checking to see if anyone was inside. I grabbed a shotgun and opened the door and then stepped out. One of the men threw up his hands but the other began talking. I recognized him right off. He was the same fellow who had appeared over two years before with two other men saying they'd been in a rollover on FM 755 north of us. He was the man who said they were walking back to a stash house in San Isidro but instead headed into the brush and then turned north to disappear into the desert. The very same *coyote* who probably had secreted spots along the way with water stored and perhaps even contacts paid to provide shelter to both him and those in his charge. "We just wanted to get some water," he said. "*¿Quién les dio permiso para entrar?*" I asked, opting for my favorite line from the movie *The Milagro Beanfield War*. He didn't reply but instead kept rambling, just as he'd done on his previous visit. "What's your excuse this time?" I asked but he looked at me confused and I imagine he didn't realize he'd been to this place before. Maybe, maybe not. I told him to get his water and be gone. Then a few days later two more men appeared in the early morning and again I stepped out on the porch, this time with a pistol in hand. "What do you want?" I asked. But, of course, I knew what they wanted and who they were and what had happened to them. "The smuggler ran off," the older of the two said as if he were spitting out acid. "Where are you from?" I asked. But he said he didn't have to say anything if he didn't want to. He looked mean and determined and I realized I needed to let him know who was boss. "Okay," I said. "We can just bury you out here and no one will know any different." He stared coldly and so I added, "It's your choice." He looked angry and determined but at last he said they were both from Honduras. Then he spit out, "Where's the town of Mission?" I pointed at the little road at the end of the driveway. "Take that road out there and keep going south until you reach a town about four miles away. Follow the paved road that goes through the town and you'll get to Mission about sixty miles away." So he asked for water and I told them both to sit down. I looked at Matthew and he grabbed a water jug then filled it and set it on the ground near them. All the while I had my pistol ready in case they made a move. Then I asked the older man, "Why do you want to go to Mission?" He replied, "Because *el coyote* lives there and I'm going to find him and kill him." Norma had made a couple of sandwiches and we placed them in a plastic bag and Matthew set the bag on the ground next to the water jug. As the older man opened the bag I said, "You'll need that food if you plan to walk that far." Later that afternoon I ran into Ray Vance who was spending a couple of days at his travel trailer along with his wife, Jovita. He said, "These two fellows walked by this morning and one of them said you'd given them food and water." Then Ray added, "The older guy kept asking how to get to Mission." So I told Ray what the man had told me. Ray was quiet a moment and then said, "Oh well. One less coyote, I guess."

As would often occur, months would go by without any activity. One day, however, near sunset the dogs began barking. I looked across the front yard and saw three people standing at the end of the driveway. An older woman dressed in black and a girl wearing a green shirt and dark blue pants. Next to them stood a young man who wore black pants and a faded olive long-sleeved shirt. They were looking at the house waiting for someone to walk out to them. So I grabbed my pistol and approached them. "Are you alone?" I asked as I scanned the woods all around to make sure there weren't others hiding.

"We're alone," the older woman said. "Can you please call the Border Patrol?"

I walked them into the yard and we sat on concrete blocks set next to a mesquite tree and when I turned I saw Norma standing on the porch holding a topped-off water jug and three plastic glasses. I walked to the porch for the jug and Norma went inside to make sandwiches. That night I wrote the following in my blog, *Woods Roamer*: "What I witness out here is genuine survival. I encounter people who have faced extreme conditions, some of them for up to a week walking in temperatures as high as 110 degrees Fahrenheit with little to no water, with no backpack full of gear, with no knife, no matches, no food, and no flashlight or ferro rod. I find them panicked and sick. I find them disoriented, gone crazy from lack of water. I find them wild-eyed and in shock. And I find what's left of them after they succumbed to the heat and their tongues swelled up to where they could no longer talk or even breathe and they collapsed on sunbaked sand and lapsed into unconsciousness, and hopefully they died quietly because when night came the coyotes, mountain lions, and wild hogs sniffed their scent and by morning all that's left is a scattering of blood-red bones. By noontime the buzzards have finished the task of sweeping away any vestige of who they might have been. The three folks who walked up to the cabin even as the dogs barked furiously were nearing the end. The older woman was suffering from acute dehydration. Her fourteen-year-old daughter was scared though in better shape than her thirty-nine-year-old mother. The twenty-seven-year-old fellow with them was weak and dazed. They were weeping and pleading for water. "Please call *la Migración*," they said. So I called the Border Patrol then asked them to sit under a mesquite tree while I brought them water and some peanut butter sandwiches. I knew it would take an hour at least for the Border Patrol to make it out to my place so we began talking about what they'd just gone through as well as about their long journey across parts of Central America then through Mexico and ultimately to the ranchlands along the South Texas desert known as the Sand Sheet.

The woman was holding a heavy black Bible that looked well worn. She said they'd carried a couple of cans of peaches and a few bottles of water but the water was long gone and so were the peaches. The woman said the smuggler had told them to rest. Exhausted from their walk they quickly fell asleep. When they awoke *el coyote* was gone. But in a twist they said there'd been two *coyotes* and both were *Americanos* with light-colored eyes and pale skin and they spoke with heavy accents. She said the men seemed familiar with the area and that one of them had walked ahead of them and had even climbed a deer tower to scout the surrounding flats looking for the Border Patrol. But in the night both men disappeared and when the group awoke they were alone.

"We walked up and down a fence line looking for houses or windmills but saw nothing," the young man said.

"We came to the same tower one of the *coyotes* had climbed the day before and we thought about spending the night there but then realized we wouldn't last another day in the desert," the mother said.

"Why do you carry a Bible?" I asked the older lady knowing that the added weight was a burden and that most long-distance travelers opted to carry water and food and nothing else.

She looked at me and shrugged but said nothing more so I asked her what religion she belonged to. "Pentecostal," she said, then added that she and her daughter were from San Salvador, the largest city in the country of El Salvador. The young man said he was from San Salvador as well but had never met the two women until they crossed the Rio Grande together three nights before. He said there'd been seventeen in the group but they'd split up after the smugglers disappeared in the night.

As I learned during my years along the Sand Sheet, most of the people who ventured north were from cities and as such unfamiliar with the ways of the woods. They knew nothing about survival or bushcraft or anything related to living in the wilderness. They came from areas of massive human concentrations and had arrived at the US border after journeying north on a train called *La Bestia*—The Beast—that plodded through Mexico laden with thousands of indigenous people looking for a way out and a way in. But the three who sat with me under the mesquite tree that evening had seen enough.

"We wanted to go to Maryland," the young girl said.

"Yes," her mother added. "We have relatives there."

"And how about you?" I asked the young man.

"I have friends and relatives living in Georgia," he said.

They're all here, I thought. By the millions they're here. They work the farms like slaves for men and woman who themselves will not work and who will not pay a living wage and who often treat them with less dignity than they do their dogs. They work the hotels and restaurants. They build houses and mow lawns.

They lay asphalt on highways and dig ditches in cities. They open taco stands along the roads or shops that fix flat tires. They sweep the floors and clean the bathrooms of large chain stores after the customers have gone. And yes, they also join gangs and peddle drugs and kill each other in their barrios and colonias. For, in fact, they are no different from others who came before them and did the same things.

So I called the Border Patrol as we sat talking about their lives in El Salvador. In some respects, I knew what they had known in their homeland would not be all that different here. Of course, they would not have understood nor would they have believed me. As it turned out Matthew was driving back from McAllen and when he reached the first gate three miles to the south he encountered a Border Patrol unit.

"Do you need to go through?" Matthew asked the agent.

"Yes, I need to pick up some illegals," the agent said.

"Well, they're at my house so just follow me in," replied Matthew.

When Matthew arrived in his pickup truck with the Border Patrol vehicle behind him I sensed the people were scared and it seemed the older woman wanted to cry.

"You'll be okay," I assured her.

"Will they separate us?" the young girl asked, looking at her mother.

"No," I said. "You'll stay together."

The three thanked me for giving them food and water and I think as well for talking to them for over an hour until the Border Patrol arrived. But I too was thankful for the visit and for their honesty about their lives and aspirations. They smiled and I smiled too, though I felt sadness knowing I'd never see them again. "Thank you," the older woman said. "You'll be okay," I repeated.

Six days later as I roamed a trail to the north I ran into three Border Patrol agents walking through the brush. Those encounters were always a bit unsettling since I never knew whether or not I was meeting experienced woodsmen or neophytes. More often than not I was encountering transplants from the city who were out of place in the woods. As such they were often nervous and perhaps even scared. They were also armed and that meant such meetings were potentially dangerous. A few years before an anxious Border Patrol agent had pulled a gun on me and almost fired a shot. It was on another piece of property I'd owned and I was walking with one of my blue heelers looking at native plants and bird watching and enjoying the quiet. I had a pair of binoculars dangling around my neck and an old lever action carbine in .357 magnum caliber secured over my shoulder with a sling. It was an area of heavy smuggling traffic and I'd spotted groups of drug runners as well as those entering the country in violation of immigration laws. As I stood admiring a red-billed pigeon in a Texas ebony tree I heard someone yell, "US Immigration. Freeze or I'll shoot." I looked to my right and saw a man pointing a pistol at me. He was so nervous he looked like he might keel over. I said, "Put the gun down." For a moment he seemed confused, so I repeated. "Put the gun down." He lowered his pistol and I said, "What the hell are you doing? This is private property." But he just stood there and didn't say a word. Then he turned around and walked off. When I drove out my property's gate he and another agent were waiting in their marked vehicle.

"Are you going to shoot me now?" I asked.

"I just want to make sure you're okay," he said.

But I was in no mood to discuss anything with him. I drove off and called Norma to tell her what had happened and that I'd nearly been shot. By the time I got home she'd already called the Border Patrol station in McAllen to complain. The man she spoke with told her he was the station chief and that many people had been hired without being properly vetted and that there were agents who had no business being in the Border Patrol. To which my wife said, "I'd better never see another one of those guys on our land again!" A couple of days later I was asked to drop by the McAllen station to talk to the agents about landowner issues, but at the time I wanted nothing to do with the Border Patrol.

On that hot afternoon as I watched the three agents walking down a trail I contemplated how I was going to get their attention without startling them. They appeared tired and bored. Not once had they looked up to see if anyone was nearby. At last I called out, "Hey Border Patrol." They stopped and looked around but it took a few seconds before they located me. "You guys okay?" I asked. Apparently realizing they were in no danger one of them said, "Just looking for another body."

I recalled what the young man had said a few days before when he and the two women asked me to call the Border Patrol. He said there'd been seventeen in their group. Was the body one of those the smugglers had abandoned? Or had he been a member of some other bunch? One of the agents said the corpse was already well on its way to becoming nothing more than bones. I looked around to see if I spotted vultures circling nearby but the sky was clear.

"There might be others," I said.

"We've been finding quite a few in Brooks and Jim Hogg," one of the agents muttered.

Both counties were a few miles north and northeast and unless people were escorted by a smuggler who had hid water along the way the chances of making it were slim. We stood there a moment and one of the agents said there were people coming in to retrieve the body. I kept thinking how it was someone's son or a father with children far away. They did not know and perhaps would never know what had happened.

"You'll find only a few," I told the agents.

But they said nothing more and I sensed that two of them were no longer there. They had left the place or at least had retreated from the scene. One of them looked into the brush nearby as if he no longer wanted any part in this business. The other just stared back down the trail not looking right or left or at me or at the others. Just staring. Not saying a word. He too wanted out, I could tell. The third agent was on the radio, and after he finished talking he asked, "You live around here?"

"Over that way," I said as I pointed south.

That night I heard a Border Patrol helicopter flying low to the northwest and I wondered, as I often did, how many people were out there on the Sand Sheet walking north. How many had died that day? How many were in the process of dying? I wondered about the smugglers. Were they nearby leading groups down sand-covered trails? How many were crossing the Rio Grande sixty-five miles to the south at that very moment? How many of those would be walking north within twenty-four hours? And what of those Border Patrol agents who were out of place in the wilds and just wanted to go home? In fact, the US Border Patrol has no real idea how many people cross or attempt to cross the Sand Sheet yearly. Apprehension rates don't amount to anything other than the number of people caught. And yet these same apprehension rates are used by politicians and by those with political agendas to claim that illegal entries are down (or up) and those same rates are often used to make policies or even pass laws. But I was told many times by Border Patrol agents that for every long-distance traveler caught as many as ten may have snuck through without being apprehended. Likewise, no one knows for sure how many people succumb to the heat and sand and lack of water on the great South Texas desert. Over a ten-year period, the number may go into the thousands. Good Samaritans have taken to leaving fifty-five-gallon plastic barrels filled with gallon water jugs along farm roads 1017 and 755. Placed on right-of-ways and marked by long poles waving a white, red-cross flag they are considered abandoned and thus anyone can open a barrel and remove a gallon jug if needed. Of course, vandals have on occasion knocked the barrels over and scattered the plastic jugs. It seems the issue of providing water to those crossing illegally has angered many people.

The helicopter circled for about thirty minutes and then flew off in the direction of the Rio Grande Valley to the southeast. Within seconds the quiet returned with only the lone whistles from a pauraque nearby. Then somewhere far off a group of coyotes began singing melancholy songs.

16

We built a small hawk tower two hundred yards or thereabouts behind the house. Matthew and I and my grandson Jacob, who'd arrived from Maryland to visit his Papa. There was so much I wanted to teach Jacob but never enough time. A two- or three-week visit once a year and then he'd be gone. When he was little, he and his father would arrive and we'd head off to the coast for a few days fishing at Port Mansfield. Then it would be time to leave and as I watched the plane flying away I'd realize I wouldn't see them for another year. For some the idea of family is a passing thing that slips away, and I wonder if they ever stop to contemplate what they missed. But when I was a young man I determined that if I ever had children they would know their father. All else in my life would be secondary. I would not leave them alone nor would I deny them my love. I tried my best but I always wonder if it was enough. I've asked them on occasion, but sometimes I think it was not the sons but their father who needed more. Now that I am old and they are gone, a loneliness creeps over me that at times seems overwhelming. Matthew's homecoming saved me and I always gave thanks that he was with us at the cabin. In time I gave him and his younger brother the land. Still, not an hour of any day went by that I did not think about my sons who lived far away. Nomar, Jason, Ethan. How will I be able to sleep knowing I let you go into the world alone, my sons?

Matthew would sometimes take his sleeping bag and a pillow and spend the night in the hawk tower. Fourteen feet high with a four-by-eight-foot floor and a wooden fence, it was a place to search the sky for hawks and vultures or look up into the clouds or watch the sun drop below the horizon or perhaps simply peer into the surrounding woods while listening to the quiet. During the night I'd glance out the window and in the starlight see that shadowy line marking the trees in the distance. I'd listen a moment and think about my son sleeping in the hawk tower hidden amid tall mesquites. Like his dad, a man of the woods. On one night a thick fog rolled in and every hour or so I'd look out the window thinking about Matthew and his brothers far away. But it wasn't until an hour or so after dawn that I spotted my son walking back to the cabin.

"How was it?"

"It was nice."

"See anything?"

"A deer a few yards away sometime after midnight. I could see it moving in and out of the starlight. Then the fog settled in."

One evening I walked to the hawk tower and when I approached I saw a pair of eyes looking down at me from the platform. I watched as a gray fox sauntered down the steps and into the brush. Atop the tower I found the fox had made himself a nice little place to spend the night. I'd sit atop the tower watching birds or on rocking chairs on my front and back porches or from my bedroom window as I worked at my computer. I continued my nighttime walks, sometimes encountering kangaroo rats hopping gingerly along microtrails cut through the grass and between shrubs. I'd sidestep spiny *mala mujer* plants that grow in abundance along the Sand Sheet.[1] One man said his brother accidentally brushed against a *mala mujer* while he was wearing Bermuda shorts. "My brother had a bad reaction and we had to take him to get treated."

Nearly five years had passed since we arrived at our tiny enclave on the sand. My world gradually became a place comprising only about three or four square miles. Everything beyond was distant and foreign. In the beginning things appeared almost insurmountable. But no, I'd been there already.

In time you forget about the sand. Not that it goes away or that you choose to ignore it but because nature has a way of hiding things if given a chance. I think of nature as a veil spread across the earth, whether covering a Mayan pyramid or discarded steel drilling pipes and wellheads or even sand swept inland from a vanished coastline. Shrubs and grasses spread over the sand and even if the rains fail, the plants remain to hide what lies beneath. Only when humans appear does the sand rise up and again take hold. Overgraze the land and the sand becomes real. Make roads and drill for gas and oil and the sand reemerges. When the winds blow the sand takes to the air and the sky becomes matted, unwashed, and disheveled. But leave nature to its own designs and the sand sleeps beneath a soft covering plied close to the earth and between solitary motts where pocket mice and kangaroo rats and deer, coyotes, javelina, raccoons, coatimundi, bobcats, weasels, skunks, ocelots, and cougars await the night.

A *loma* is a hill and a *lomita* is a little hill. There are no *lomas* on the Sand Sheet but instead *lomitas*. Perhaps the more appropriate term is simply to call them sand dunes sometimes mixed with a bit of clay. The closer one gets to the Gulf Coast the more dunes one observes. But in South Texas they are called *lomas* though they should really be called *lomitas*. Regardless whether one chooses *loma* or *lomita* or sand dune, these masses of sand inch across the land north by northwest in a lemming's journey to oblivion. For even a *lomita* will eventually succumb to nature's will. And so on this present-day savannah that in 1834 the French botanist, Jean Louis Berlandier, called a "wilderness of plains," I built a house.

The esoterica of science finds good lodging within the Sand Sheet with *deflation troughs* (temporary ponds) and dune descriptions from *banner* to *stabilized*. But for the people living along this wilderness of plains it is simply home. And at last, the Sand Sheet had become my home. When I was away I missed my little place. I missed the nights listening to pauraques whistling and great horned owls hooting. I missed coyotes singing in the distance. I missed sunsets that seemed as if painted across the heavens by an artist who used every color available from his palette. At last the Sand Sheet had become mine and upon that acceptance I saw two Sand Sheets. I saw the Sand Sheet that men choose to exploit and the Sand Sheet that strives to survive. I pursued the latter. Walking trails with earnest searching for that world that lives even as humankind plays its own lemming game. Finding a quiet place and then sitting and waiting for the stillness to settle upon me. The Sand Sheet can only be known by those who choose not to isolate themselves from it; whether standing on an oil derrick or riding a tractor or ensconced within a little house atop a tower they will never know what surrounds them unless they step out and turn off the machines and find a place to sit and listen.

On cool autumn nights and even cold winter nights I worked in my little shed making knives and selfbows and whatever else I fancied. A warm glow emanating outward from beneath the sheet metal roof. My blue heelers reposed on the concrete blocks that Amador laid to form the floor. Now and then I stop to listen to coyotes close by or a screech owl trilling from the nearest mott. On the Sand Sheet time is marked by other things. Mockingbirds begin chirping at one in the morning. A half moon rises an hour later. In the interim a mesquite branch evolves into a ladle as I work a hatchet to form its rough shape and then take a gouge and hook knife to make the bowl. A tillering stick is fastened to one of the shed's support posts and I stand back to examine my latest bow. It needs a little more work on one limb and then the arch will be perfect. A farriers rasp and then a couple of Nicholson rasps, numbers 49 and 50, and then a fine mill file and finally a scraper worked back and forth shaving paper-thin strips off the stave. A man and his beloved dogs with a new bow in his hands. At midnight I put my tools away and sweep the floor. Flip off the light switch and walk to the house. I remove my boots on the front porch and set them on an archery target. I put on my slippers and follow the walk-around to the utility room where I take off my clothes and set them in the washing machine. I'll wash them tomorrow. Wearing nothing but my slippers I walk back to the porch and linger a moment in the frigid air looking out into the blackness. A few years back I was dying. And now I'm standing butt naked on the front porch of a little cabin in the woods. Thank You.

Notes

CHAPTER 2

1. Texas State Historical Association, *Texas Almanac*: "Origins of the Camino Real in Texas." http://texasalmanac.com/topics/history/origins-camino-real-texas.

2. Joachim A. McGraw, "Archaeology in the South Texas Sand Sheet: A Study of Chevron Properties in Brooks County." *Index of Texas Archaeology: Open Access Gray Literature from the Lone Star State*, vol. 1984, article 7.

3. Martín Salinas, *Indians of the Rio Grande Delta* (Austin: University of Texas Press, 1990), 164.

4. Antonio Zavaleta, professor of anthropology, University of Texas–Brownsville, personal communication.

5. The proper definition of a *ranch* is a homestead and not an expanse of land. Throughout Mexico and South Texas people refer to their rural homesteads as *ranches*. That is the correct meaning of the word.

CHAPTER 12

1. The word *tiller*, as applied to bows, comes not from the words "sprout" or "twig" but from the concept of "steering," i.e., any method or mechanism that promotes the proper steering of something. To tiller a bow therefore refers to the act of shaving the belly side of a stave in order to attain an arc when the bow is pulled to final draw length. Properly tillered bows will thus steer an arrow with greater accuracy since energy is stored equally on both sides of the bow's handle and that energy is transferred equally into the arrow when the string is released.

2. A selfbow is made of one piece of wood without laminations or backing like sinew, rawhide, or linen. Selfbows are the oldest form of bow making requiring the greatest knowledge of wood structure and properties. Although more challenging to make than other bows like laminated or backed bows, the selfbow offers the greatest satisfaction for the bowyer.

3. Martín Salinas, *Indians of the Rio Grande Delta* (Austin: University of Texas Press, 1990).

4. Ibid., 126

5. The Coahuiltecan Geographical Region encompasses most of what is known as Deep South Texas and northeastern Mexico. The various Neolithic family groups and bands that lived within this area were linked by social and linguistic commonalities but were not part of any greater tribe or political assemblage.

CHAPTER 14

1. Coahuila is a Nahuatl word that makes up part of the Uto-Aztecan languages. http://www.omniglot.com/writing/nahuatl.htm.

2. EARTHWORKS Hydraulic Fracturing 101, http://www.earthworksaction.org/issues/detail/hydraulic_fracturing_101#.VLv9b0fF-8A.

3. At this writing the oil and gas industries continue to deny their culpability in earthquakes related to deep disposal wells in North Texas and Oklahoma.

4. Spine weight refers to arrow flexibility, and an "archer's paradox" is the phenomena that occurs when an arrow must negotiate its path around a bow's grip and then (assuming correct spine weight) flex and continue in the direction initiated by the string's travel.

CHAPTER 16

1. *Mala mujer* (bad woman) is also known as Texas bull nettle. http://www.wildflower.org/plants/result.php?id_plant=CNTE.

GLOSSARY

1. Common names and scientific names of plants are found at the US Department of Agriculture *Plants Database*, http://plants.usda.gov/java/, and the Texas A&M System *Plants of Texas Rangelands Virtual Herbarium*, http://essmextension.tamu.edu/plants/.

Glossary

Common and scientific names of the plants mentioned in the text:[1]

Agave (*Agave americana*)
Anacua (*Ehretia anacua*)
Ashe Juniper (*Juniperus ashei*)
Barretta (*Helietta parvifolia*)
Brasil, Capul, (*Condalia hookeri*)
Caña (*Arundo donax*)
Carrizo (*Phragmites australis berlandieri*)
Cattail (*Typha* sp.)
Cedar Elm (*Ulmus crassifolia*)
Chaparro Prieto (*Vachellia rigidula* or *Acacia rigidula*)
Chapote (*Diospyros texana*)
Colima (*Zanthoxylum fagara*)
Granjeno (*Celtis pallida* Torr.)
Huisache (*Vachellia farnesiana* or *Acacia farnesiana*)
Live Oak (*Quercus virginiana* var. *fusiformis*)
Lotebush (*Ziziphus obtusifolia*)
Mala Mujer (*Cnidoscolus texanus*)
Mesquite (*Prosopis glandulosa* Torr.)
Ortegia (common nettle) (*Urtica dioica*)
Osage Orange (*Maclura pomifera*)
Pecan (*Carya illinoinensis*)
Retama (*Parkinsonia aculeata*)
Texas Ebony (*Ebenopsis ebano*)
Texas Lantana (*Lantana horrida* renamed *Lantana urticoides*)
Wright Acacia (*Acacia greggii* var. *wrightii*)
Yucca (*Yucca treculeana*)

Index

adze, 85
agave cordage, 31, 67, 94
Algonquin tribes, 41
Amador, 19, 33, 34, 35
amargosa, 82
anacua stave, 33, 98
ashe juniper, 98
asphaltum, 94
atlatl darts, 91
atlatl points, 30

badger holes, 87
Baffin Bay, 23
barranca, 40
barretta, 93
Berlandier, Jean Louis, 138
blind snakes, 48
las brasadas, 22
brasil, 6, 12, 13, 20, 27, 39, 81, 83, 85
Brooks County, 3, 23, 29, 130
Buddy, the bobwhite quail, 67, 68, 69
Burns, Steven, 97
bushcraft, 65

caña/*Arundo donax*, 31
canícula, 57
capul, 78
carrizo/carrizo arrows, 31, 91, 94, 95, 97, 119
cattail arrows, 113
Central Market, 103, 104
Chagas' disease, 96
chaparro prieto, 31, 85, 93, 98
chapote, 27, 63, 88, 98
chauite, 95
chile del monte, 88
Ciudad Camargo, Tamaulipas, 121
Ciudad Mier, 121
Coahuiltecan Geographical Region, 1, 95, 97, 109, 113, 114
colima, 13, 39
Colombian exchange, 95, 96
coma, 82, 88
Comanche traces, 22, 114
Corpus Christi Bay, 23
crooked knife, 30, 31, 39, 40, 41, 84, 85

cougar, 57
croton, 24
curandero, 84

D-bow design, 114
Delmita, TX, 29
desert yaupon, 82

Eagle Ford Shale Region, 66, 107
ebony, 81, 85, 93, 98, 129
El Centro, TX, 6, 29, 95
el desierto, 13
endoscopic retrograde cholangiopancreatogram (ERCP), 7
eolian drifts, 13

ferrocerium rod, 30
Fort Sam Houston, 104
fracking (hydraulic fracturing), 65, 66, 112, 113

Goliad Formation/Goliad soil, 14, 119
Gonzalez, Juan, 119, 120
granjeno, 6, 12, 13, 19, 27, 39, 41, 57, 78, 81, 83, 88, 93, 114, 122
guayacan, 82, 85

haboob, 23
Henry, the screech owl, 69
Hernandez, Mario, 62
Hidalgo County, 3, 12
Holocene, 1
hook knife, 84, 85, 86
Hoyt, Ruth, 68
huisache, 85, 98

jacales, 97
Jim Wells County, 3
Jim Hogg County, 3, 130

Kenedy County, 3
kingsnake, 48
Kleberg County, 3

La Bestia, 127
Ladrón de Guevara, 93, 94, 95, 97

La Gloria, TX, 29, 107, 112
lantana, 13
La Reforma, TX, 29
leather stem, also dragon's blood, *sangre de drago, Jatropha dioica*, 81, 82, 83, 84
lippia, 56
loma/lomas/lomita, 137
Longoria, Ethan, 12, 24, 135
Longoria, Jacob, 135
Longoria, Jason, 12, 53, 135
Longoria, Matthew, 12, 24, 103, 105, 119, 120, 122, 124, 128, 135, 136
Longoria, Nomar, 12, 53, 135
Longoria, Norma, 12, 24, 25, 33, 47, 53, 71, 72, 73, 74, 75, 104, 124, 125, 129
Lopez, Andres, 122
Lopez, Joseph, 121, 122, 123
Lopez, Omar, 106
Lopez, Tololo, 33, 47, 106
Los Caminos Real, 22, 114
lotebush, 13, 39

mala mujer, 136
Mara Salvatrucha, 71
MELD Score (Model for End-Stage Liver Disease), 7
mesquite, 6, 12, 13, 19, 25, 27, 31, 42, 56, 67, 78, 83, 85, 87, 93, 97, 98
mesquite sap glue, 95
mono-scape, 13
motts, 12, 13, 27, 41, 81, 99
musette bag, 27, 31, 55, 86

Neolithic people/Neolithic natives, 1, 115
nopal, 77
northern cat-eyed snake, 48
Nueces River, 115
Nuestra Señora de la Santa Muerte, 73
Nuevo León, 23, 109
Nuevo Santander, 23

oak, 13, 83
orquetas, 31
ortegia cordage, 31
Osage orange, 98

palo blanco tree, 7, 58
parang, 30
pauraques, 14
pecan, 98

Peña, Ramiro, 28, 29
Peña, Santiago, 28
percutaneous transhepatic cholangiography (PTC), 7
Phragmites australis berlandieri/ Phragmites australis americanus/ Phragmites australis pinolios, 95, 96, 97, 114, 119
Pleistocene, 23

Rachal, 112
ramadero, 29, 88
retama, 122
Rockwell (measure), 86
Roman Catholic Church, 22, 93

Saenz de Guerra, Ines, 29
Saint Isidore of Seville, 28
Salinas, Martín, 93, 94
salvia, 56
San Fernando/San Fernando River, 81, 93, 94, 97
San Isidro, TX, 6, 15, 21, 25, 33, 47, 61, 74, 75, 76, 77, 106
Santa Elena, TX, 29, 76
selfbow, 1, 31, 33, 45, 66, 91, 92, 116, 138
Showronek, Russell, 119
Speeg, MD, PhD, Kermit, 8, 9, 10, 66, 99, 103
Starr County, 3, 12, 19, 71, 93, 95

Tamaulipas, 23, 81, 93, 109
transition zone, 81, 82, 83, 84, 87, 120
Treviño, Benito, 48, 121
Treviño, Toni, 121
tsunami of sand, 82, 83

University of the Incarnate Word, 104

Vance, Ray and Jovita, 34, 124

wilderness of plains, 138
Willacy County, 3
Woods Roamer blog, 45, 49, 65, 111, 125
Wright's acacia, 85

yucca cordage, 31

Zapata County, 3, 12